Earth-Saving
Acts
— FOR —
Eco-*Warriors*

**JOIN THE FIGHT FOR
A SUSTAINABLE
FUTURE**

STERLING
New York

CONTENTS

WELCOME

Our planet is facing tough environmental challenges right now. Oceans are filling up with plastic, rainforests are being destroyed, and wildlife is suffering. With all this doom and gloom surrounding climate change, is there really anything you can do to make a difference?

The answer is yes. Today, there is a growing movement of eco-warriors who are rallying against the unsustainable exploitation of the natural world and the authorities that refuse to listen, and you can be one of them. It's about making responsible choices and inspiring others to join you. Taking action can begin with something as small as using a refillable cup. Even turning lights off when they're not being used can have an effect. Or why not find out more about the biodiversity you have on your doorstep? Are there ways you could help your community to become more aware of any issues?

Filled with ideas and advice on everything from cycling to upcycling, we hope this beautiful guide will give you a better understanding of what is really happening to the world around us, and empower you to take those small steps that will lead to positive change for the future. Because together we can all make a difference.

WHAT DOES IT MEAN TO BE AN ECO-WARRIOR?

Get the lowdown on environmental activism.

What does it mean to be an eco-warrior? A whole host of things. Being an eco-warrior means you care about the environment and the world you live in. It shows you love the creatures that you share the Earth with and know they are precious. It demonstrates you appreciate the beauty of the sea and countryside, the mountains and jungles. Being an eco-warrior means you understand the importance of looking after our planet and you want future generations to be able to enjoy everything that's wonderful about it, too.

It means you're prepared to stand up for a green sustainable lifestyle—one where you embrace eco-friendly practices. You want to recycle items that can be repurposed, reuse items that would otherwise go to waste, and reduce the amount of natural resources consumed. Being an eco-warrior means you encourage everyone—including yourself—to make responsible choices, every day.

Living a green life should be a priority for everyone, regardless of age, where they live, or what they do. The survival of the planet is dependent on people joining forces and unitedly making a difference. Climate change should not be ignored. With Australian wildfires engulfing millions of acres of land with devastating consequences, summer temperatures in European countries hitting an all-time high, and Atlantic hurricanes getting stronger and intensifying more quickly, it's never been more evident that the world is warming up.

It's an alarming predicament, and people are responsible for it happening. Homes are using more energy from fossil fuels, such as oil, gas, and coal, than ever before, emitting dangerously high levels of carbon dioxide into the atmosphere. Humans are destroying forests too, with the equivalent of twenty-seven football fields being cut down every minute for agricultural and commercial use, again contributing to high levels of greenhouse gases. All this is to blame for global warming.

Making an effort to live sustainably can help to save our planet—and the people, animals, and plants living on it—but only if everyone acts quickly. If glaciers and ice sheets continue to melt, where will polar bears build their dens and raise their young? If sea levels rise further, what will that mean for coastal towns and their residents? If rain forests keep being destroyed and animal habitats threatened, how can certain species survive? And if water pollution causes further damage to the world's coral reefs, what will happen to other marine life?

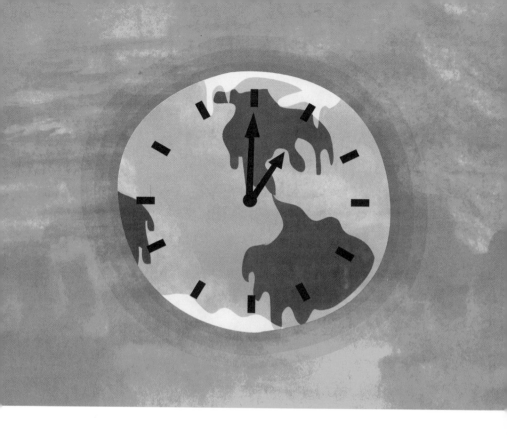

Time to change

Governments around the world, all very aware of the climate change emergency, came together in 2015 to start discussions and later signed the Paris Climate Agreement. The deal's ultimate goal is to substantially reduce global greenhouse gas emissions in an effort to limit the world's temperature increase in this century to 2°C above preindustrial levels. It was a historic turning point, and 197 countries claimed to be committed.

Fast-forward a few years and Finland, India, and Lithuania are all making progress with renewable energy; Morocco and Gambia are both on track to cut their carbon output; and Costa Rica has announced its plans to be the world's first carbon-neutral country by 2050. Worryingly, however, Russia, Turkey, and Iran have still to formally join the agreement, while America is set to fall short of its Paris pledge and has signaled its intent to withdraw.

There is hope though. One of the world's most powerful voices on climate change is teenager Greta Thunberg from Sweden. Frustrated that no one appeared to be taking immediate action to combat climate change, she skipped school in August 2018 to protest outside the Swedish parliament. "I can't vote, so this is one of the ways I can make my voice heard," Greta explained at the time.

Her commitment to tackling global warming started an international youth movement, with tens of thousands of students around the world choosing to protest, too. While you may not have Greta's high profile (well, not at the moment!), remember she's just a young person with a passion, someone who realized it was important to be green, acted on her strong beliefs, and shared her concerns with others. You, too, have the power to make a difference. You can live in a green way. You can motivate family and friends to be eco-aware. You can help save the planet.

As Greta has publicly said: "The climate crisis has already been solved. We already have all the facts and solutions. All we have to do is wake up and change."

WHAT IS A CARBON FOOTPRINT?

Find out all about yours, and what you can do to reduce it.

Did you know that around a fifth of your body is made of carbon? It's an element found in every living thing on the planet and when combined with oxygen it forms an invisible gas called carbon dioxide, or CO_2. This greenhouse gas traps heat and energy that the Earth gets from the sun. Without CO_2, all this energy would leak back out into space.

So, what's the big deal about CO_2 emissions in our atmosphere if the gas plays an important role? Well, today too much CO_2 is being produced, which is causing the Earth to become warmer. We're experiencing a man-made climate change and if it continues at its current alarming rate, the future of our planet is at risk.

Traveling by car, charging a cell phone, eating a beef burger—everything we do creates CO_2. Humans and animals, of course, breathe out CO_2, however the problem is escalating now because in our modern world more and more fossil fuels are being burned—things such as petroleum, coal, and natural gas that were all formed in the earth from animal or plant remains. This process produces a significant amount of CO_2 emissions.

Reduce your impact
When people talk about a carbon footprint, they're referring to how much an individual's life affects the environment. It's not visible like a footprint in the sand, but everyone has one. The energy you use at home contributes to your footprint— the food you eat and clothes you buy have also, at some point along the line, released CO_2 emissions and so add to your carbon footprint, too.

The good news is you can help to decrease the amount of CO_2 in our atmosphere just by making some wise everyday choices. Understandably, you do need to use a computer for work projects, you will watch TV, and your home is a bus ride away from a gym, but it is still possible to be carbon neutral and balance everything you do that releases carbon emissions with a carbon saving elsewhere.

Want to start reducing your carbon footprint today? Here's how:

TRAVEL

With cars, buses, trains, and airplanes all dependant on fuel, it's no surprise that the travel sector is the main culprit of global carbon emissions, responsible for nearly one tenth of greenhouse gases. Every time you drive somewhere in the car or you jump on a train, this adds to your carbon footprint.

The greenest way to get from A to B is on foot, but sometimes walking isn't really an option. Could you cycle instead? Traveling by bicycle is around ten times more carbon-efficient than the same journey in a car.

As for flying, the carbon footprint of a return flight from London to New York is more than what a person in Paraguay will generate in a year. Going on a plane to a vacation destination may be fun, but you can also have a great time on a break closer to home, and you'd be helping to save the planet.

What you can do:

◆ Get on your bike
Cycle as much as possible instead of using a car. Your carbon footprint will be lower, and it's a healthy way to keep fit.

◆ Use public transport
If you're traveling a distance, again avoid using a car. Take a bus or train, and share the carbon footprint of the journey with other passengers.

◆ Enjoy a staycation
Instead of traveling across country or vacationing abroad, take a break in your home state. If you are intent on going farther afield, forgo a flight and look at alternate ways to travel.

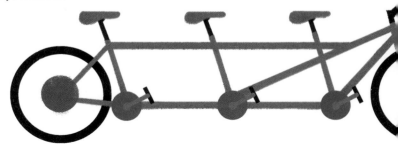

FOOD

Food also has an environmental footprint, equating to a quarter of global emissions. Greenhouse gases are produced by growing, rearing, farming, processing, transporting, storing, and cooking all the things you eat. Meat, particularly lamb and beef, is the worst offender—cows also release methane when they burp or pass wind—but eating dairy products, poultry, fish, seafood, and eggs will also add to your carbon footprint.

Buying local produce and growing your own vegetables reduces food miles—the distance products have to travel to a supermarket. Choosing organic seasonal food is environmentally friendly as well. However, regardless of what ingredients you use to make meals, households should try to reduce their food waste. One third of all food produced ends up in the trash, and annually in the US wasted food creates the same amount of CO_2 emissions as 37 million cars.

What you can do:

⊙ **Swap meat for vegetables**
You don't need to become a vegetarian or vegan, but you can reduce the number of meat-based meals you eat. You'll find plenty of great recipes that use vegetables, beans, and nuts.

⊙ **Shop locally**
Encourage your family to shop locally and buy from local producers so fewer food miles are involved. Plan dishes that use seasonal ingredients.

⊙ **Cut down on takeout**
Eating takeout food that is packaged and delivered to your home increases your carbon footprint, so prepare food at home instead.

ENERGY

Whether it's using a hairdryer daily or charging a smartphone, homes across the world use an incredible amount of energy from fossil fuels, which creates CO_2 emissions. Although you may not be responsible for which company provides electricity for your home, families can cut their carbon footprint by switching to an energy supplier that produces electricity from the wind and sun. Wearing a sweater when it's cold instead of turning up the thermostat can reduce your carbon footprint too.

Using energy-saving lightbulbs can reduce the CO_2 emissions that are released into the atmosphere by around 20 percent. And when households need to replace a washing machine or refrigerator, they can purchase energy-efficient appliances. Do note though that upgrading an electronic device adds to your carbon footprint, so think twice before asking for a new cell phone or iPad. Apple claims 80 percent of a new laptop's footprint comes from manufacturing and distributing the product, not using it at home.

What you can do:

⊙ Switch off lights and gadgets
Turn off the light when you leave a room. And pull out the plug when an electrical device is not in use. Don't leave a computer or TV on standby either—they still suck up energy.

⊙ Take a shorter shower
Heating water uses energy, so don't spend all morning in the shower. Turn off the tap when brushing your teeth, too.

⊙ Don't fill the kettle
If making a cup of tea, coffee, or hot chocolate, don't fill the pot or kettle to the brim. To save energy, only boil as much water as you need.

SMALL WAYS TO MAKE A BIG DIFFERENCE

More things you can start doing today to reduce your carbon footprint:

* Carry a reusable shopping bag.
* Draw on both sides of paper.
* Shop at thrift stores, or swap clothes with friends.
* Reuse gift bags or wrapping paper.
* Close the curtains to keep a room cool, avoid air conditioning.
* Always use a recyclable water bottle, not a plastic one.
* Eat fewer packaged snacks.
* Play outside games, not video games.
* Encourage others to reduce their carbon footprint too.

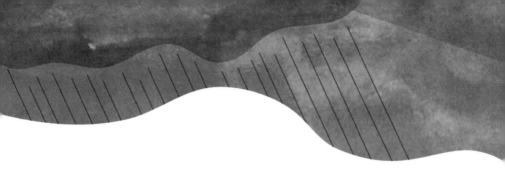

DO YOU HAVE ECO-ANXIETY?

Are you feeling overwhelmed by the challenge of climate change? Is the state of our oceans keeping you awake at night? If so, you're not alone.

"Climate catastrophe," "12 years to save the planet," and "Irreversible damage" are just a few examples of the scary-sounding headlines bombarding us almost every day about our changing planet. It's hardly surprising, then, that with these and the numerous concerning stories about the devastation caused by plastic, deforestation, and global warming, many feel worried about the future.

Some people aren't just worried, however—they're overwhelmed by the stories, and it's affecting their mental health. According to the American Psychological Association, eco-anxiety is the "chronic fear of environmental doom" and it's being felt by many people, but particularly the younger generation who are concerned about their future and feel powerless to stop the damage.

Anxiety is a condition that affects many people of all ages, with symptoms including panic attacks, tightness in the chest, and a sense of dread and fear. Those with eco-anxiety may feel some of these, as well as a sense of loss, hopelessness, and frustration that they're unable to do anything about climate change. Their sleep may be affected, and they may develop obsessive thoughts. Others may worry about how hurricanes, tsunamis, and other disasters they've heard of or seen on TV could affect them in the future.

WAYS TO COPE WITH ECO-ANXIETY

⊙ **Speak to someone**
The most important first step if you're feeling anxious or overwhelmed is to talk to an adult, whether it's a family member, friend, teacher, doctor, or nurse. Tell them about your concerns for the environment but, more importantly, explain that it's affecting your mental health. Don't be embarrassed or feel your worries are too silly—lots of people feel the same, and it's crucial you talk about them so you can live a happier life.

⊙ **Be calm and breathe**
Find a way to help calm your mind when you're feeling stressed, such as a mindfulness app or using breathing techniques to stop the rush of thoughts. It's important to address the anxiety and look for ways to cope as well as being proactive in your efforts to improve the environment. Anxiety is common, and it doesn't mean you're crazy. It's important to know you can overcome it, no matter how out of control it feels.

⊙ **Find your tribe**
Being involved can make you feel more positive and, with the help of others, you will feel as though you're having more of an impact. Join a local environmental group or find out about the protests for climate change that take place each month. By surrounding yourself with like-minded people, you're more likely to feel confident about the changes that can be made.

❯ Identify triggers

Is there a particular environmental problem that makes you feel anxious? If so, work out how you can face that fear. Research the truth of the matter. Some damage may be irreversible, but there are still plenty of things that you can do to help ensure a brighter future. Climate change is a long process, and scientists are looking for solutions, such as electric cars, solar panels, and recyclable products. No one can change the past, but people are trying to have a positive impact. Be one of them!

❯ Take action

You may only be one person, but you can make a difference by changing your lifestyle and reducing your carbon footprint. Consider how much electricity you use, where you get your food and clothes, and how you travel. This will make you feel more empowered and less stuck in your anxiety. No one can live the perfect "green" life, but small changes do add up.

❯ Inspire others

It can be frustrating to see people dropping litter without considering its impact, or friends who don't seem to care about the products they buy as much as you do. Why not organize an event for your family and friends where people do something to make a difference? This could include recycling or swapping clothes, collecting trash, or having a meat-free day. You could even offer to give a talk at your college or workplace and help other students (and teachers) or colleagues to see how actions taken now can affect everyone's future.

NEWSFLASH: IT'S OUR TIME

Student and political activist Ruby Etwaria Sweetman explains why it's important to get involved in politics at all levels, so the issues that matter to you can't be ignored by the people in power.

Who takes the garbage out in your house? Your parents, a friend, an older sibling, you? Perhaps you take it in turns. What you can't determine or change is the day on which it's collected—that power lies in someone else's hands and they form part of your local government. They might seem removed from the more pressing aspects of your everyday life (garbage-day is hardly a big deal), but these people also make bigger decisions—ones that will affect your life and career as well as the future of the planet.

Think about where you are now, for instance. If you're at college, it's the people in power, the politicians, who decide which subjects are allowed to be taught, how they're taught, the length of courses, and how much can be

charged for tuition fees, and what's paid back in student loans. That's why it's important to stay in tune with and understand new rules and laws they suggest. After all, they'll affect your world, and you might not even agree with them. It isn't easy, however, to make room for politics and current events when you already have so many other things going on.

So, why should politics be part of your life? Because it affects all our lives, whether we like it or not. And politicians speak on our behalf, so it's important that we elect the ones who speak for us. Understanding the events that influence political decisions also provides a more rounded view of the world, which is valuable in further education and the workplace.

In February 2019, I took part in a Friday climate-change walkout, inspired by climate activist Greta Thunberg, along with other friends concerned about the health of the planet. It felt good to be part of a group that was taking action rather than watching from the sidelines. It also felt powerful to be with like-minded people who were all there for a common goal, demanding change from the current politics. These opportunities and events shape our future, and we will soon be in charge.

WAYS TO GET INVOLVED

▷ **Where to start?** You could begin by researching different political parties and their leaders and see if any have views that are similar to your own.

▷ **Passionate about an issue?** Join an organization or a political party.

▷ **Listen and learn.** Tune in to a range of talks and podcasts, or read introductions to a certain subject.

▷ **Establish good habits.** If you're just beginning to get interested in politics and the world around you try watching the nightly news or reading newspapers online.

▷ **Find your own niche.** Feel like politics doesn't affect you? Think of an area of your life you're passionate about and explore how it's run and who pays for it. For example, if you're an artist, funding for the arts is a political issue; if you're interested in reading, plans for your local library are determined by politics and money.

▷ **Be curious.** Pick an area of your life you care about deeply and research what the world thinks about it. Adopting an inquiring mindset will help your political awakening. If you approach news with a curious mind—rather than deciding on your opinion before you've heard all the facts and weighed up all the options— you'll be more open to debate and discourse.

▷ **Be smart about your sources.** News is complicated, and sometimes the truth of a story can be hidden, making it hard to pick out reliable sources from the less trustworthy. And remember that news, whether it's on the TV, in a newspaper, on a blog, on Instagram, or BuzzFeed, isn't neutral. The same story can be represented in many ways depending on the writer's perspective.

BEWARE FAKE NEWS

It isn't always easy to tell what's real and what's fake, but these tips might help:

* Check your source. Are you reading or watching a reliable news platform?
* Never rely on only one source for information. Check the details against at least two other reliable sites.
* Who has published the information? Is this news, a rumor, or is it just outright gossip?
* What is the context of this post? Is there more you need to find out to understand fully its meaning and purpose?
* Explore what other news outlets are saying about the event (looking at other countries' news platforms can be interesting) as well as seeking the opinions of friends and family members.

PEDAL POWER

Getting on your bicycle isn't just eco-friendly, it's good for you, too.

Can you remember that magical moment when you could cycle for the first time with no help? When you're really young, it's great to get out on your bicycle, but as various other pastimes and activities kick in, many give up this inexpensive (after the initial cost of the bike and safety gear, that is) pleasure. But that bike collecting dust in your garage is more than a metal frame with two wheels—it can become your new best friend, accompanying you on thrill-seeking rides or when you just want to escape for some fresh air, get somewhere on time, or visit a friend.

Get into gear

One of the best things about cycling is how easy it is to get started. You just need a bicycle, a helmet, and a willingness to pedal, whatever your level of fitness.

Even if you haven't cycled before, if you're committed, it's entirely feasible to build up to riding quite long distances within a couple of months. Just start with a couple of rides a week and gradually increase the distance. If you haven't ridden a bike in a while, consider cycling somewhere traffic-free on your first few outings, perhaps in your local park or dedicated cycle path, then research some quiet routes to try while you familiarize yourself with riding on the roads.

Physical benefits

Cycling is also a great way to keep fit and a brilliant way to tone up muscles around your glutes and legs. Whether you're just getting into fitness or want a new exercise challenge, cycling helps you whatever level you're at because you can choose your route, the level of gears, and how hard you pedal.

As well as boosting your physical health, cycling will also improve your mental health. Why? Firstly, exercise releases feel-good chemicals in your body. It will also give you a feeling of accomplishment as your fitness improves and you develop a new skill. You'll get a boost from the fresh air, too.

Going solo

If you're confident enough, cycling alone can be great for your soul and an excellent self-esteem booster. It allows you freedom to exercise at your own pace, time to process any stresses or anxieties playing on your mind, and space to come to decisions you might need to make. You also get time to savor things you don't notice if you're in a car, such as the animals in fields, the smell of hot chocolate from a kiosk, people you pass on the way, and beautiful buildings and landmarks.

Social cycles

Some of the best times can be had on a group cycle with friends or family—not only does it give you a chance to bond or catch up, but it also means you can be more adventurous with routes, woods, and trails. If your family and friends aren't into cycling, there are many clubs around that specialize in everything from fun social rides to competitions in velodromes, road racing, and mountain trails. Who knows, you could be a future Olympic cycling star! But the most important thing is just to enjoy the ride.

SAFE CYCLING

If you're going to take up cycling, here are some tips to keep you safe:

⊗ **Check your bicycle is road ready.** Have you got pumped-up tires, oil on the chain, and working brakes? If you're unsure how to check, or how to fix a puncture if one were to happen, ask to be shown guides online.

⊗ **Tell others.** Tell a friend or family member where you're going before you venture out solo. Remember to stay in safe areas and take a phone with you.

⊗ **Be prepared.** Ensure you're ready with a repair kit, water, and food just in case. Glasses can help to protect your eyes from insects and debris that might fly up from the wheels. Are you familiar with road safety, signals, and road awareness? Check the safety guides from the National Highway Traffic Safety Administration in the US and Canada's Ministry of Transportation: nhtsa.gov/road-safety and mto.gov.on.ca/english/safety/bicycle-safety.shtml.

⊗ **Be aware.** Wherever you're cycling, you need to be alert at all times, so avoid listening to music and be mindful of walkers, fellow cyclists, and drivers.

⊗ **Be safe.** Wear a properly fitted helmet. You may be a safe rider, but you can't do anything about the behavior of other road users, which is why it's also important to be seen. Reflective clothes and good working lights and reflectors on your bike are essential in low light and sensible at any time of day.

For more information on cycling, choosing a bicycle, or joining a local cycle club, visit the American Bicycling Education Association at abea.bike or the Canadian Cycling Association at cyclingcanada.ca.

FACT OR FICTION?

The world is awash with misleading advice about how to help save the planet—we separate the myths from the truth.

Switching your gadgets off, using eco-friendly products, and becoming a vegetarian are among the many suggestions given to those wanting a greener lifestyle. But are you being given the correct advice? Or are there more to these tips than it seems? We take a closer look at ten widely held green views.

If an appliance is off, it's not using electricity

This is wrong. Even when appliances are switched off, some still use power. Known as "phantom power," it applies to items on a standby mode such as TVs, refrigerators, computers, phone chargers, or items that display a clock, like microwaves. The truth is "off" is not enough—unplug your devices instead.

Green cars are the answer

Electric cars use less fuel and produce much lower emissions than regular cars, so they are the greener choice. However, making electric cars requires more energy and emits more CO_2 emissions than building a normal car. These cars redeem themselves during their lifetimes with their low emissions, but cycling is still the greenest option.

Organic is always best

When you see the word "organic" on a product, you think it is better for you as it's produced without chemicals and using environmentally friendly methods. But look at where your products have come from before considering how green they are. Often they're shipped from the other side of the globe and have used up a lot of fuel to get to you, meaning harmful CO_2 emissions. Yes, organic is good, but locally grown produce is another way to be eco-friendly and local organic produce is even better.

We should stop using palm oil products

The devastation caused to wildlife by the palm oil industry has been widely publicized over the past few years. However, with the oil featuring in almost 50 percent of products, it's hard to avoid buying it. In fact, not buying products with palm oil could force companies to grow products that are even worse for the environment and take away support for those trying to make the palm oil industry more sustainable, such as the Roundtable on Sustainable Palm Oil (RSPO). So next time you are shopping, look for the RSPO label and check the palm oil included is sustainable. If you find a product that you like contains unsustainable palm oil, write to the company that produces it and ask them to take action to make a change.

Paper bags are better than plastic

Some people believe paper bags should replace plastic ones in stores as they're made from products that can be recycled and are also biodegradable. However, paper bags require trees to be cut down, and making them results in fifty times more water pollutants. Of course, plastic is still a problem. But replacing them with paper bags isn't the easy answer. The key to reducing the impact of all carrier bags is to reuse them as much as possible.

Vegetarian diets are better for the environment

Animal products are responsible for a huge amount of emissions because of the resources required to raise, feed, and transport them. It's also widely known that a cow's burps and farts are full of methane, also a greenhouse gas. However, dairy products that use a lot of milk can have more of a carbon footprint than some meats. The most eco-friendly option is to go vegan instead.

Washing in cold water won't get your clothes clean

Heating water uses a lot of energy, but some people believe only hot water will get rid of tough stains. In fact, washing your clothes in warm or even cold water gets rid of most dirt. And switching from hot to warm water can cut your energy use in half.

Overfilling a kettle wastes energy

This is true, but it's not just the amount of water boiled that counts. It's the kettle chosen. Plug-in varieties are convenient but use more energy than burning gas on a stove would to heat a pot. If you want to reduce emissions, switch to the stove, especially in the colder months when the heat from the flames will also help to heat the room.

A warmer planet is a good thing, surely?

For some, yes, warmer is more beneficial as crops will grow better and in countries that don't see sun often, hot days are huge events. But on balance, there's a lot of suffering as a result and the growing number of heatwaves is a cause for concern. Since 1906, the global average surface temperature has risen by 1.6°F (0.9°C) which has been melting glaciers, leading to rising sea levels, droughts, and a range of problems for wildlife and landscapes.

It's too late to do anything

Climate change is a huge problem, and unless big changes are made, the world will continue to be irreversibly damaged. However, it'd be too easy to throw our hands up in the air and give in. Everyone can ensure that the most disastrous predictions won't come true by making changes now to reduce their carbon footprint and protect nature for future generations.

THE PROBLEM WITH FAST FASHION

Are you always trying to keep up with the latest clothes trends? If so, you could be harming the environment without even realizing it.

Fashion's impact on climate change is one of the biggest talking points in environment discussions at the moment. This is because the fashion industry is a major source of the greenhouse gases that are overheating the world. A whopping 1.3 billion tons of carbon emissions are released annually—more than international flights and the shipping industry combined—as a result.

To make matters worse, according to reports a staggering 75 percent of the 58 million tons of fibers used to make clothes and textiles each year are burned or sent to landfill. In short, the textile industry has become polluting and wasteful, which is why environmental campaigners are urging people to reduce their carbon footprint by opting for more sustainable fashion.

This refers to clothing that is environmentally friendly because of how it was designed, produced, and marketed, and how it can still be worn and reworn many times in the future. It sounds a fantastic idea in theory, but sadly the fashion industry isn't sustainable and our clothes choices continue to impact our environment.

WHAT'S THE BIG DEAL?

◉ People love fast fashion

It's trendy and cheap (much more so than sustainable clothes) so it's no wonder it's popular. But often the clothes are made in one country, manufactured in another, and sold worldwide to meet fashion trends, impacting the environment due to the corners being cut to make them so quickly and cheaply.

◉ Cheap implications

To make cheap clothes, you need cheap material and cheap labor. Low-priced clothes are often made in factories abroad, and there is a high risk that the workers have been paid low wages and suffered poor working conditions. In these factories, which use a lot of energy to meet demand, shortcuts can be made that affect the environment, such as water pollution or noxious gases being released.

◉ Material matters

Every time you wash a piece of clothing made from synthetic materials like polyester or nylon, you are contributing to plastic pollution. Around 700,000 man-made fibers are released in a normal wash, which flow into the sea and are eaten by fish. Clothes release around half a million tons of microfibers into the ocean every year, equivalent to more than fifty billion plastic bottles.

◉ Landfill issues

So many people fail to recycle their unwanted clothes and the equivalent of one truckload full is wasted every second. Campaigners are urging people to repair, recycle, or resell clothes more to help reduce the amount of waste. Materials often won't break down in landfill and are there for decades, failing to rot away.

HOW CAN YOU MAKE A DIFFERENCE?

Shop smart!

While it's tempting to buy new clothes, don't feel under pressure to conform to the latest trends. Take time to think about what suits you, what you will wear regularly and items that will last. Why not follow the #30wears campaign, which encourages you to only buy something if you know you'll wear it at least thirty times? If you do buy, choose fabrics wisely to minimize your impact on the environment.

Do your research

Find out more about where your clothes come from by looking at the label or researching the company on the internet to see how much they reveal. Check out the material, too. Cotton in particular isn't as environmentally friendly as you might think—it can require a high level of water and pesticides to be made—so see if yours is organic. Sustainable fashion may be more expensive, but it will last longer and is better for the planet.

Go thrifting

Instead of buying new clothes all the time, check out your local thrift stores and find a bargain. Not only is it a fun experience as you never know what you'll find, but you can buy good-quality items cheaply and raise money for others at the same time, depending on the store.

Make your clothes last longer

Even if you don't want your clothes anymore, don't throw them out. You can increase the life of your items by either donating them to charity, taking them to a recycling bank, or swapping them with friends and family. If something is broken or ripped, take time to mend it rather than throwing it out.

BLOCK PARTY

Discover how you can make a difference in your neighborhood.

Getting involved in your community is one of the best ways to keep eco-anxiety at bay, and with climate change and green concerns being the leading issues of the moment, a surprisingly high number of your neighbors are likely to share your concerns. Here are some easy and fun suggestions for you to encourage your community to come together and make changes for a sustainable tomorrow:

Gather your eco-tribe

Do you have an idea for an event, activity, or resource that could make a real difference in your community? Perhaps you'd like to organize upcycling craft sessions, an annual plastic-free festival, or a sponsored fundraising event? The best way to turn an idea into reality is to have the support of others who also like your idea. And while online groups have their place, you can't beat a meeting IRL. So first things first, go on a scouting mission around the neighborhood—perhaps taking a friend for support (and a sensible safety precaution)—to get a rough idea of how many people are potentially interested. Then do some research to find a suitable venue—a local church or school hall, for example. Start with a monthly meetup to discuss plans. Don't be shy. Discuss your ideas with friends and neighbors—you might even be able to get local media coverage about it.

Throw a "swap, borrow, or buy" garden party

Buying new shiny things might give you a short-term buzz, but the more eco-savvy choice is to use what you and your neighbors already have. It's time to start loving on the old stuff, so ask everyone to bring the items they're keen to swap, loan, or buy. The scooter, sneakers, or games console you have outgrown can all have a happy afterlife with their next owner. And the feel-good buzz you'll get from reusing old goods is priceless.

"Green up" the neighborhood
Plants are pretty incredible, like nature's own superheroes. Their eco-credentials include cleaning pollutants from the air, producing oxygen, and providing homes for vital wildlife. And they look awesome, too. Community projects like the American Community Gardening Association can offer advice and support if you're ready to try "greening up" your neighborhood, visit communitygarden.org.

Organize a clean-up timetable
It's time to talk trash with your neighbors—removing garbage from rivers, for example, can help make them safer habitats for aquatic life and improve water quality for people, too.

Host a talent show
Suggest an eco "show-and-tell" night, where neighbors demonstrate their environmentally friendly life hacks. Whether it's transforming an old T-shirt into a fashionable wardrobe addition or sharing a passion for DIY beauty products, everyone is sure to learn something new and, most importantly, have fun! Pinterest has lots of inspiring ideas, such as homemade bath salts and seriously cute paper beads made from—drum roll, please—cereal boxes.

⟫ Cook up a storm

When you consider that food makes up around a quarter of your greenhouse gas footprint, it's worth finding ways to reduce it. An easy idea? At the eco-neighborhood get-together, ask everyone to bring a favorite dish (making sure that you don't end up with ten bowls of pasta salad!) to share. Large quantities of food use less packaging and less energy to cook. Cutting back on meat and dairy is also a way to take less from the environment.

⟫ March like you mean it

Grass-roots environmental groups like FridaysForFuture (fridaysforfuture.org) are helping to give people a platform to radically reshape the political landscape around climate change. But if you get a little freaked out by big organized rallies, consider having your own local march instead. A get-together with your neighbors to make placards and banners is also a great way to find out more about each other and discuss new ideas. You could even use the time to draft emails to lobby your local government representative with your eco-concerns.

Volunteer

Many community groups rely on volunteers, whether it's fundraising for a cause, helping at coffee mornings, running events, or managing day-to-day activities. Perhaps you could give some of your free time and energy to help a local environmental group, animal shelter, or after-work eco club?

Share your gifts

Are you good at art? Can you sing or play a musical instrument? Are you a computer wizard? Offer your talents to the community. You can use your gifts for fundraiser events or to help raise awareness of important issues.

STEPS TO SUCCESS

The first step to changing your community for the better is simply to decide that you want to have a positive impact. But how can you ensure that any changes you make will be effective and long lasting? Here are four tips to keep in mind when you're just starting out:

Do your research
Once you've chosen a problem in your community that you'd like to focus on, find out if anything has already been done to try to solve it. If so, where did previous efforts succeed and where did they fail? Were they missing something that you could provide? Knowing the answers to these questions may help you to avoid common mistakes.

Put your plan on paper
With your research out of the way, make a plan of action, and write it down. Studies show that written goals are far more likely to be achieved than non-written ones.

Measure your progress
Now that you have a realistic goal in mind, it's time to ask yourself some questions: Why is this goal important to you? How often will you volunteer your time toward it? How much do you hope to achieve in a month? Or in a year? When will you know it's been a success?

And remember . . .
Most ways to change the world depend on support—even if you have life-changing ideas, you can't do it alone. Whatever your idea, find like-minded people who are willing to help. And never underestimate the power of changing one person's life. Every act for your community adds up and can lead to lasting change.

YOU'VE GOT THE POWER!

Everything you need to know about renewable energy—and how you can make a difference.

If you want to discuss climate change and how to make a positive difference, you need to make sure you know your facts about renewable energy, as it's one of the hottest topics when it comes to saving the planet and preventing global warming. So, what is it and why is it so important?

What is it?
Renewable energy is a resource that can be used repeatedly because it's replaced naturally, such as water, wind, or solar energy. While much progress has been made in finding ways to use renewable energy, most of the globe is still being powered by nonrenewable sources such as coal, gas, and oil. These are fossil fuels burned by power stations to generate electricity for homes and transport.

The main problems with fossil fuels are that they're running out quickly and also that when they're burned, they emit CO_2 and other greenhouse gases that trap heat in the Earth's atmosphere, causing average temperatures across the globe to rise. As a result of this global warming, there have been extreme weather shifts such as heat waves, floods, fires, rising seas, and other negative impacts.

Renewable energy—known as green power or clean energy—therefore, is a key weapon in the fight against global warming. Turn the page to discover some examples of these amazing resources.

Wind power
A breeze can be converted into electricity using wind turbines. The wind rotates the turbine blades, the energy of which produces electricity via a generator. Wind energy generated 6.5 percent of US electricity in 2018, enough to power 26 million homes.

Solar energy
This technology converts the sun's rays into electricity using solar cells on panels. You've probably seen many examples of their use from panels on roofs to whole fields full to car parking meters.

Hydro power
Like with wind, water can spin turbines and generate electricity. Hydropower is currently the world's biggest source of renewable energy by far, with the US, Canada, China, Brazil, and Russia being the top producers.

Wave and tidal power
The energy of the ocean can be harnessed and turned into a power source, again by spinning turbines.

Geothermal energy
The heat from deep underground can be used to make electricity. It can produce steam which goes up a pipe and then spins a turbine, powers a generator, or goes directly into pipes to heat homes.

Biomass energy
This uses organic material—or natural products—to generate electricity. For example, sawdust, leftover parts of certain plants and trees, waste, and manure can all be used to create energy.

RENEWABLE
ENERGY FACTS

* Just one wind turbine can generate enough electricity to power over 300 homes at the same time.
* If taken advantage of to its fullest extent, one hour of sunlight could meet the world's energy demands for an entire year.
* According to the World Wildlife Fund (WWF), the whole world could get all its energy from renewable resources by 2050, but only if the right political, financial, and society decisions are made quickly.
* One-third of the world's installed electricity generating capacity is from renewable energy sources.
* Two-thirds of the power capacity added around the world in 2018 was from renewable energy.

So, why isn't everyone using renewable energy?

How much you use depends on which country you live in, how much they've invested in these energy sources, and who your supplier is. In Iceland, for example, a huge 100 percent of their electricity comes from renewable energy, in comparison to just 17 percent in the US. Most countries, however, have a mix of fossil fuels and renewable sources that can be accessed by energy suppliers. Although it seems the best solution, establishing renewable energy sources is often expensive, and ideal locations for turbines or systems need to be found.

Green energy confusion

Some people still incorrectly believe renewable energy is more expensive, which is another reason why more people aren't switched on. Others think swapping to clean energy means attaching your power to a wind turbine. Again, this isn't true. There are many green energy companies that supply as much electricity from renewable energy sources as possible and invest in sourcing green energy for the future.

What can you do to help?

* Consider your own digital carbon footprint and how your choices emit CO_2 through the amount of electricity generated by your TV watching, phone scrolling, and Internet use.
* Every video streamed, search made, and song played uses electricity. It's estimated that the energy used for the Internet accounts for 4 percent of greenhouse gas emissions and this figure is set to rise each year.
* Unplug gadgets and electronic devices and switch off lights that aren't needed.
* Sign up for Earth Hour at earthhour.org and join the individuals, landmarks, and businesses who all turn off their lights in solidarity for the planet.
* Can't wait for Earth Hour? Why not challenge yourself and your friends to an hour without electricity to see how it goes?

Young activists who are bossing it around the world:

* Delaney Anne Reynolds, from Miami, helped build a solar-power charging station for cell phones at her school.
* When Malawi-teen William Kamkwamba was fourteen, he built an electricity-producing windmill from spare parts and scrap, working from rough plans he found in a library book. A book about him, *The Boy Who Harnessed The Wind*, has recently been turned into a movie.
* Jamie Margolin cofounded the protest group Zero Hour in Seattle in 2017, when she was just fifteen. Today, she is bringing American teens together to help stop climate change.

TO FLY OR NOT TO FLY?

The Swedish anti-flying movement "flygskam" is gathering pace across Europe, but what is the philosophy behind it and can it make a similar difference in other parts of the world?

"Do you need to fly—can't you take a train?" If you've been on the receiving end of a comment like this after excitedly telling your friends about your holiday plans, you're not alone. Making people feel embarrassed for flying is part of a growing movement called flygskam, a Swedish phrase that translates as "flight shame."

Flygskam was born after singer Staffan Lindberg wrote an article encouraging people to give up flying for environmental reasons in 2017. His personal stand, and the idea of advocating more eco-friendly travel to others, was supported by fellow Swede Greta Thunberg, the teenage climate change activist who, rather than fly, crossed the Atlantic in a racing yacht to take part in the United Nations Summit in August 2019.

It's important to note here that the word "shame" is misleading. Flygskam advocates aren't suggesting everyone should be booed as they get off a plane. But they do feel that people need to become more aware of their own choices. The point of the movement is to turn off the autopilot (pun intended) when it comes to booking flights, instead asking yourself: "Is there another way to get there—could I get a train, bus, boat, or even cycle instead? And if not, should I go somewhere else?"

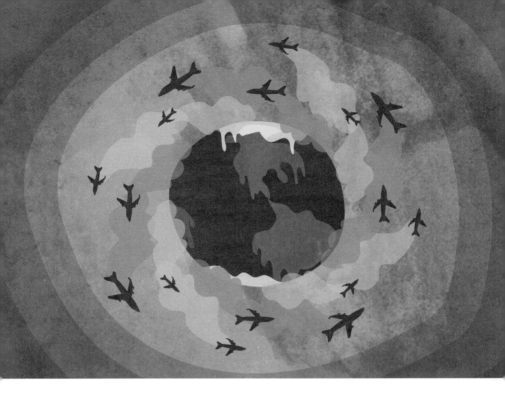

But is flygskam really necessary?

"Please fasten your seatbelts and return your tray table to its full upright and locked position." How many times have you heard this instruction while sitting on a plane? Chances are you've heard it a lot—especially if you're American or British. In 2018, British people flew abroad more than any other nationality, taking a staggering 126.2 million flights, equating to roughly one in twelve of all international travelers. Americans were close behind, leaving the country via plane a huge 111.5 million times in just a year. More shocking still, US airlines carried 777.9 million passengers on domestic flights across America in 2018.

Flying is still one of the world's most popular forms of travel, yet it's also one of the most polluting. A study by *The Guardian* found that one long-haul flight generates more carbon emissions than an average person does in a whole year. Even a relatively short return trip from New York to Miami carries a carbon footprint of over 515 pounds (234 kg) of CO_2 per passenger, which is more than the average produced by a person annually in seventeen countries.

But what difference can one person make?
Lots! A survey of more than 6,000 people in July by Swiss bank UBS found that one in five travelers is flying less as a direct result of "flight shaming." The respondents, from the US, Britain, France, and Germany, said they had reduced their air travel by at least one flight in the past year because of climate concerns. If these trends continue, the expected growth in passenger numbers could be halved, the report said. Fewer passengers means fewer flights, which means fewer emissions.

Scandinavian airline SAS AB saw a 2 percent decrease in all flights this year, while Sweden said it had dealt with 9 percent fewer passengers for domestic flights. Companies are also jumping onboard, with Swedish bank Klarna Bank AB banning all employee air travel within Europe and discouraging long-haul flights.

Flygskam is having a real and tangible effect, making people conscious about their own responsibility to the planet. But if you do have to fly, there are a few things you can do to make your flight more eco-friendly:

⊙ **Purchase carbon offsets:** Many airlines now have an option for paying to offset the emissions of your flight. You just pay an extra fee on top of the flight cost, which is donated to a carbon offset scheme. Ask the airline about it when you book your flight.

⊙ **Choose an environmentally-conscious airline:** Pick one that has a newer fleet of planes (which are lighter and burn less kerosene), uses more eco-friendly materials, limits the use of single-use plastic, has paperless cabins, and serves healthy meals.

⊙ **Always choose a direct flight:** Stopovers can increase a flight's total emissions by up to 35 percent.

⊙ **Pack light:** If everyone aimed to pack 2 pounds (0.9 kg) less than normal, it would be equivalent to removing more than 10,000 cars from the road annually.

⊙ **Fly economy:** Reducing space per passenger means less emissions per passenger. Bring your own reusable bottle for drinks—and why not bring your own snacks too, to reduce waste?

It becomes an ethical choice: If you can choose alternate means of travel, why wouldn't you? And you won't just be helping the planet: a brilliant and unforeseen benefit of flygskam is the way it changes how you view journeys.

Small change, big impact

There are numerous mantras and mottos about learning to "savor the journey" within life, rather than just focusing on the destination. The point being that too often people set their sights on a goal and don't stop to enjoy the process of achieving it. In choosing not to fly and having to find alternative routes to wherever you want to end up, the journey itself becomes part of the adventure. Instead of jumping on a plane and zoning out for a few hours, you can watch breathtaking scenery from a train window, marvel at the size of the ocean while on a ferry, or play a game of cards with a stranger on a bus.

Do what you can. It's near-enough impossible to have zero carbon impact but small changes, such as avoiding domestic flights, can make a big difference—and sharing what you're doing in a positive way will encourage others to do the same. Search #flygskam on Instagram or Twitter and feel inspired by all the amazing posts from people who've discovered new places, new routes, and new friends through choosing to go flight-free. Look into what your government is doing regarding the issue. Are they following the Swedish government's plans for an eco-tax on aviation? If not, why not?

Flygskam is changing the way people think about the ethical, practical, and emotional ways they travel—and there's no shame in that.

TREE OF LIFE

*From gathering the seeds to caring for the young
sapling, a decision to grow your own tree can bring
rewards for a heap of species—including humans.
Here's how to get started.*

Growing a tree from seed might not seem like it will change the world for the
better, but with care, the right conditions, and a bit of luck, your seedling could
grow to become a magnificent specimen that'll benefit future generations.

Why more are needed

Every year, countless trees are lost as a result of disease, pollution, and old age,
but trees are also a valuable resource and many are felled to provide everything
from food and paper to furniture and fuel. Some of this commercial felling has
resulted in deforestation, where large areas of forest are cleared and not replaced.
This can cause a collapse in natural habitat and threaten native species. Trees
also support endless insects, birds, butterflies, and, of course, humans, by
absorbing CO_2 from the air.

Choose your variety

Take a look around your local park or countryside and you'll notice that
there are many different trees. Now, using a guide (you can find
one online) see if you can discover any of the following, which
are common to the US: maple, pine, Douglas fir, aspen, or oak.
Keep in mind that some trees (like the oak) can take decades
to mature, whereas others, including weeping willow,
Lombardy poplar, and silver birch, grow quickly.

GROW YOUR OWN

What you'll need to get started:

* A few tree seeds
* Plant pots with drainage holes
* Good organic compost
* Small stones or pebbles

▶ Collect and prepare seeds

Seed collecting takes place in the fall, around September and October. If you're not sure where and how to collect them, ask your local conservation trust for advice or take part in a seed-gathering event. You might be familiar with the ones from oak (acorns) and maple (propeller seeds), but you'll find tree seeds come in all different shapes and sizes. There are nuts and cones, as well as pods, wings, and fruits (apples, berries, cherries) where the seeds need to be extracted. Remember that tree fruit and seeds also offer food for wildlife, so don't take more than you need. Once you've gathered enough, keep them cool and dry until you're ready to sow them.

▶ Sow the seeds

Place a few stones or pebbles at the bottom of a plant pot and then fill it with compost. Create a hole in the soil, about 1-inch (2.5 cm) deep, and plant the seed. Cover with compost, lightly press down, and then water it lightly. Put the pot in a warm, dimly lit location and check it every couple of weeks, watering only to ensure the soil doesn't dry out. You'll now need to be patient. Most tree seeds sown in the fall tend to germinate the following spring, but it can take even longer. In some cases, germination can fail, so you might want to sow a few seeds in different pots to increase the chances of success.

▶ Nurture

As the seed sprouts, ensure that the soil is kept moist (but not overly wet) and that the seedling has enough light and air. Move the pot to a brighter place so that sunlight encourages it to grow. Keep an eye on it and, if it's outdoors, protect your sapling as best you can from pests and wildlife. Repot the tree as it grows. Once it reaches about 16 inches (40 cm) in height, it's ready to be transferred to the ground.

Where to plant your tree

If you don't have a yard that's big enough to accommodate the eventual height and width of a tree, you could give your sapling to a community woodland. Contact your local nature conservation society or wildlife trust to find out about tree-planting projects in your area. Once your sapling is planted in a suitable spot, it will still need care and protection from wildlife, especially in rural areas where deer might be roaming. But as each year passes, your young tree will put down deeper roots, become stronger, and grow to its full potential.

More information about trees is available from The Arbor Day Foundation. Visit arborday.org for lots of great tips.

GENTLY DOES IT

Some people relish the idea of being an activist and shouting to be heard in a bustling march or protest, while others find the idea scary and a bit overwhelming. If you're in the second camp, you may prefer a quieter kind of activism, one where you use creativity and individuality to get your message across.

When you think of an eco-activist, what images spring to mind? Noisy, crowded marches with placards? Greta Thunberg addressing the world's most powerful and influential politicians? Passionate people shouting about their cause? Maybe even people doing extreme things, such as chaining themselves to trees or sabotaging an event, to get their point across? With so many issues affecting our world, sometimes there is a need to "be loud" to get people to sit up and listen, but it isn't always the best way to get people to take notice. Plus, you may not like that style or feel too shy to consider that type of campaign. That doesn't mean you can't be an activist and promote your cause, however, because there's a new style of campaigning that's growing in popularity—it's called gentle activism.

WHAT IS "GENTLE ACTIVISM"?

No one likes being shouted at or feeling bullied into agreeing with the ideas of others. While preaching and protesting can work, gentle activism is a softer, quieter approach to suggesting a change in someone else's views. It encourages people by intriguing them rather than forcing them to do anything. Gentle activism is also preferential for those who aren't ready to march but are still keen to promote their cause. The key thing to remember is that to change someone's view or convince them of an idea, there are stages—first they need to be made aware of an issue, then they have to alter their mindset based on what they know, and finally they work out how to make changes. You can encourage people to do this in a range of ways.

Here are some ideas to get you started:

Use social media
Sharing good articles, TV shows, podcasts, or books on social media is a positive way to encourage others to learn about the causes that you feel are important. By just sharing and not thrusting your ideas onto someone, you're quietly but confidently educating others. Some people may comment after reading or watching, and it may encourage a good debate. So next time you read an article or see an image that resonates with you, why not share it? You may discover some of your friends feel the same but haven't been brave enough to do anything.

Do it together
Could you set up a group to discuss things that you've seen or read? Finding like-minded people and sharing ways you can become more environmentally friendly will help make a bigger difference than doing things alone. If there's a group of you, you're likely to spread the word quicker and make more of an impact.

⊘ Art and crafts
Art can have a huge impact on people and can demonstrate your feelings about something. You could make something to convey a message, whether it's knitting, pottery, a painting, video, poem, or even a song or rap. You could also make a T-shirt with an activist slogan on it to promote your ideas as you live your day-to-day life. A magazine for family and friends is another artistic way to campaign for your causes and voice your opinion. Do something to make others think and make the most of your creative talents at the same time.

⊘ Organize events
If there's a cause you're passionate about or you have an idea that you want to share so that you can make a difference, take the time to organize an event. You could run it at college or in your local area to raise awareness or educate others on how they could make changes, such as by running a clothes swap, an upcycling event, a repair café, or even a movie-screening to encourage others to get involved. Often, people know about an issue but don't know how to start, so give them a place to begin.

⬡ Show off

Persuade people that change is easier than they think by revealing snippets of how you make a difference—for example, if you're vegan and want to persuade others to do the same, invite friends and family over for delicious plant-based dinners, or share images of amazing food you've cooked or are about to eat on social media. If you want to encourage others to wear more sustainable clothing, show off about your favorite eco-clothing brand, online or to a friend. Everyone looks to others for inspiration and for some, that inspiration could be you.

⬡ Listen to others

Gentle activism isn't about keeping quiet. It's about encouraging others to make changes by setting an example or sharing ideas about how to make a difference. Being able to voice your opinion is important—but so is listening to others. Some people may not agree with you and you may not agree with them, but try not to get angry as this leads to conflict. Accept that some people may not hold the same viewpoint, however hard you may try to reason with them. Instead, focus on the changes you can make, because you can really make a difference.

NEW GENERATION

What do you think the latest generation will be known and remembered for? And should older generations be paying them more attention?

You've probably heard of terms like Millennials and Generation X. They're used to categorize people who were born during certain periods of time and who are thought to share particular characteristics. But who can exactly say when a generation begins and ends? It's in no way an exact science, but rather it consists of an age group sharing a common history and experience that reflect the influences of the time. So, what's shaping the most recent generation? And why is it already becoming a powerful and promising one?

Move over Millennials

Generations follow but rarely resemble each other. Improved technologies, an ever-more connected world, and changing economic conditions are, for example, factors that make this teenage generation different from the previous one. Millennials (those who reached adulthood around the year 2000) have enjoyed many educational freedoms and job opportunities opened up by the internet revolution, which brought the rest of the world within easier reach. They've played a big part in changing the way people shop, travel, and even how they think about work. They took the world by storm and made their mark on it, but a new generation is arriving on the scene.

The smartphone generation

Those born between the mid-1990s and around 2012 are known as Generation Z or the iGeneration. Having had lots of powerful technology at their fingertips almost from birth has tended to encourage open-mindedness and a broader view of the planet and its many cultures. There's great potential for them to use technology to positive effect and show the world what they're made of.

Shapers of the future

A digital voice is one that can be heard, regardless of its age. Views and opinions can now reach a wider audience than ever before, and it's looking like some older people are beginning to recognize that for a society to thrive, it needs to listen to the voices of its younger members. So, whether your passion or concern is climate change, gender equality, or social justice, embrace your inner activist, and get your voice heard. You have the power to stand up and fight for your beliefs, regardless of where you live, your age, gender, ethnicity, or education. Find your platform and share your message. Every voice matters.

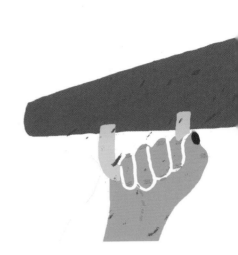

YOUNG ACTIVISTS DEMANDING ACTION

❯ For gender equality

Malala Yousafzai grew up in a village in northwestern Pakistan. She became a symbol for the right of girls to be educated when she defied the Taliban militant group that had taken charge of her community and sometimes banned girls from attending school. She was only eleven when she started writing a blog (under a made-up name) about her experiences. This caught the public eye, and in 2012 a member of the Taliban shot and tried to kill her as she made her way home on a bus after sitting an exam (two other girls were also shot in the assassination attempt). Even after this attack, she boldly continued her campaign, notably with a passionate speech to the United Nations on her sixteenth birthday. She even became the youngest person to receive the Nobel Peace Prize at only seventeen.

❯ For community

Inspired by his youth-worker mom, Jeremiah Emmanuel from London, England, started working within his local community at the age of four. He now campaigns around several issues that affect young people, including violent crime. In 2013, when he was deputy young mayor of the Lambeth city district, he founded an organization called 1BC (One Big Community) after a friend was murdered, and he has since worked at improving the links between young people from disadvantaged areas and business. In 2017, he was named on the Queen's New Year's Honor List and awarded a British Empire Medal.

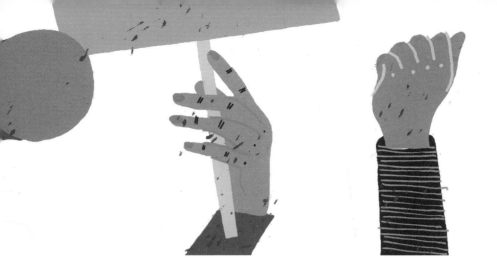

❯ Against violence

American Emma Gonzalez made one of her most impactful statements with silence. Following a mass shooting at her school in Florida in February 2018, she and a group of students organized "March For Our Lives," an event that demanded an end to gun violence. She gave a powerful speech for what she said and, memorably, didn't say. This remarkable moment would be reported to be "the loudest silence in the history of U.S. social protest."

❯ For freedom

Afghanistan-born Sonita Alizadeh has spoken up against forced marriages. Each year, twelve million girls around the world are married before the age of eighteen, and Sonita almost became one of them. When she was sixteen, she wrote about the fate of many of her peers in a rap song and, with the help of a filmmaker, created a music video, *Brides For Sale*, which went viral. In 2018, the MTV Europe Music Awards honored her with a Generation Change Award.

❯ Against climate change

Swedish teenager Greta Thunberg is an activist trying to slow down climate change. Following heat waves and wildfires in her country in 2018, she started the first school strike and inspired other students across the globe to do the same. She has since addressed the United Nations Climate Change Summit and the World Economic Forum. News magazine *Time* named her one of the world's twenty-five most influential teenagers of 2018.

SPINNING OUT OF CONTROL

Microscopic plastic particles from our laundry are threatening the planet's wildlife and, ultimately, us. So, what can we do about it?

Most people wear synthetic fabrics, such as polyester, acrylic, nylon, or Lycra, in some form every day. Dresses, pants, yoga pants, and fleeces are all increasingly made of synthetic materials containing polymers. But these man-made fabrics, from which more than 60 percent of all the world's clothing is derived, have a nasty surprise up their sleeves: When they're washed, they release tiny plastic particles—microfibers—that flow, by the billions, with the waste water down the drain, then through water-treatment plants and out into rivers, lakes, and oceans.

"Different types of fabrics can have very different levels of emissions," says Richard Thompson, professor of marine biology at Plymouth University in the UK, whose team conducted a year-long investigation into what happens when synthetic materials are washed in a machine. They found that acrylic shed the most particles, releasing a staggering 730,000 microfibers per wash.

Too small to be captured by standard washing-machine filters, the particles track through people's wastewater into sewage treatment plants where the majority slip through the filtration process and find their way into rivers, lakes, and oceans. There, they gather other pollutants to them, becoming tiny toxic time bombs, ingested by marine life, entering the food chain and, eventually, people.

Time to act

While more research needs to be done, scientists have already proven that the problem is widespread. "Clothing fibers are the most abundant form of waste material that we find in habitats worldwide, and the problem is worsening," says Dr. Mark Anthony Browne, a senior research associate at the University of New South Wales, Australia.

Unlike the plastic bag that can be photographed enticing a turtle that mistakenly thinks it's eating a jellyfish, or the deadly discarded fishing gear seen ensnaring a seal, the danger lurking in laundry is invisible. Until now, it has also largely been a case of out of sight, out of mind.

"We've known about this problem for seven years and yet the clothing industry hasn't done anything to mitigate the damage their products are doing," says Stiv J. Wilson, who works with environmental project The Story of Stuff. "They are passing the responsibility on to consumers and washing-machine manufacturers, but from an enforcement standpoint, it would be extremely impractical to filter washing-machine effluent—there are more than 100 million machines in the US alone."

Looking for a solution

German washing machine manufacturer, Miele, is one of the companies exploring the possibility of filtering out the pollutants during the wash cycle, but, as a spokesperson for the company explains, there is no simple solution: "Our product developers found that plastic matter from clothing is far too small to be mechanically filtered out, so a chemical method of removing these particles must be developed. Given the current technical situation, a solution is unfortunately unlikely to be forthcoming in the foreseeable future."

On a more positive note, the Amsterdam-based Plastic Soup Foundation, which campaigns against plastic waste in water, has launched Ocean Clean Wash, bringing together a group of scientists, industry organizations, and fashion brands

to look for solutions to the microfiber problem. "There are coalitions around the world working on how to tackle the issue, but their deadlines are far away and put the responsibility on the consumers," says project leader Laura Díaz Sánchez. "Solutions need to be developed in the short-term while long-term actions are put in place. We're not aware of any fabrics being developed that shed no fibers at all, but there's a pectin coating that can reduce shedding by 80 percent. What we need is funding into further research and development."

The full treatment
As clothing brands and washing machine manufacturers work toward effective solutions, the final link in the microfiber chain—the waste-water treatment plant—is also coming under pressure to reduce emissions. "Treatment facilities aren't currently equipped to capture these tiny particles," says Dr. Geoff Brighty, technical director for the Plastic Oceans Foundation. "Water companies are developing techniques now, but improvements in fiber capture are unlikely to come into play until 2020–25."

Meanwhile, billions more fibers will be entering the waterways and finding their way into the food chain. "Progress does seem slow, but it's a complex problem and effective solutions won't come overnight," says Dr. Brighty.

Working with the Norfolk Rivers Trust, Dr. Brighty is pioneering an innovative method of removing toxins from treated water, establishing wetlands or "aquatic gardens" which naturally remove polluting phosphates, nitrates, and ammonia. Current studies suggest that they could be effective in capturing microfibers, too.

"Most countries are now aware of the issues and taking action," he says. "We've banned polluting microbeads, and legislation is coming in to limit single-use plastics. I'm confident we will solve the microfiber problem, but by jumping too fast to find a solution, humans can sometimes do the wrong thing. Our dependency on plastics is proof of that."

SIMPLE WAYS TO CLEAN UP YOUR LAUNDRY

❯ Buy fewer, higher quality, natural pieces of clothing
At the moment, there's no evidence that natural materials shed less than synthetic ones, but fibers from natural fabrics, such as cotton, bamboo, or hemp, are more degradable, so they have less environmental impact (when not chemically treated).

❯ Use the washing machine less often
Washing a full load of laundry each time means there is less friction between clothes, so fewer fibers are released into the drum during the wash cycle.

❯ Switch to washing liquid instead of powder
The "scrub" function of the powder grains loosens clothing fibers more than liquid detergent. Where possible, opt to use refillable containers for your liquid.

❯ Wash at a low temperature
Washing clothes at a high temperature can damage some fabrics, leading to an increase in fibers being released.

❯ Avoid long wash cycles
A long period of washing causes more friction between fabrics, which leads to more tearing of the fibers.

◆ Spin clothes at low revs

Higher revolutions increase the friction between the clothes, so fibers are more likely to be released.

◆ Invest in a Cora Ball

Pop one in your washing machine and this innovative capture ball, developed by the Rozalia Project in the US, will reduce the number of microfibers in the wastewater heading down your drain.

◆ Get a Guppyfriend

Try washing your clothes inside this simple mesh bag from German outdoor brand Langbrett. It's placed inside the washing machine, capturing many of the microfibers released during the wash cycle.

◆ Clean up your kitchen

Washing-up cloths, sponges, and scouring pads can be a source of microfibers too, so try to opt for natural products whenever you can.

MAKE YOUR VOICE HEARD

*   Write an email to a handful of your favorite clothing brands or stores and ask what plans they have in place to reduce their microfiber contribution and use of toxic chemicals during the production process.
*   Contact the manufacturer of your washing machine to see what it is doing to reduce microfiber release during the wash cycle. Let it know that your next purchase will take its environmental credentials into account.
*   Encourage your government to follow Australia's lead and label the polymers used to make synthetic clothes (and other products) as environmental pollutants, making it impossible for manufacturers to continue producing materials that are known to shed plastic-derived microfibers.

SAYING NO TO A PLASTIC SEA

With so much plastic waste flowing into the sea every day, two incredible women—filmmaker Jo Ruxton and champion freediver Tanya Streeter—are working to save the world's oceans.

In the waters around Sardinia, Italy, turtles swallow bottle caps, plastic bags, and balloons, mistaking them for their natural diet of jellyfish, with deadly consequences. In the Indian Ocean, off the coast of Sri Lanka, a young pygmy blue whale rises up for air from under the waves, emerging into a mess of plastic debris and oil floating on the surface. These shocking scenes were recorded for *A Plastic Ocean*, a remarkable documentary released in 2016. It reveals the effects of plastic on the planet's marine ecosystems and human health, and launched a campaign to rid the world's oceans of plastic waste. Praised by naturalist Sir David Attenborough as "one of the most important films of our time," the documentary is the work of filmmaker Jo Ruxton, a former producer with the BBC's Natural History Unit and member of the *Blue Planet* team. "People watch wildlife documentaries and think the oceans are still pristine but they aren't," says Jo. "I've known film crews spend two hours clearing up beaches before they can take shots of turtles."

Pacific garbage patch

Through her work with the BBC and seven years as a marine scientist, Jo had seen close up the impact that people have had on the planet's seas. But it was after she had the opportunity to join an expedition to the legendary Great Pacific garbage patch—a vast ocean gathering of plastic and debris—that she realized she had to make the film. Jo was prepared to be shocked, expecting to see an enormous mass of waste, but instead she was met by what looked like a beautiful unspoiled sea. It was only by trawling with plankton nets that they revealed the water was actually choked with millions of tiny fragments of plastic, swirling with the currents and saturating the ocean. "I thought plastic was an issue of pollution, eyesore, and entanglement. I had no idea that it was a threat to human health too," she says.

Microplastic

More than eight million tons of plastics end up in the world's seas each year, with some dumped by ships, but most coming from the land and taking up to twenty years to reach the center of the ocean. "During that time, sunlight, waves, and salt make the plastic brittle, so it breaks up into tiny fragments," explains Jo. These pieces of microplastic leach out toxins, but they're also magnets for other harmful chemicals, which combine in what Jo calls "poison pills." Mixed in with plankton—the foodstuff of a huge array of sea creatures from shrimp to whales and the basis of life in the oceans—the particles end up in the food chain. Inevitably they make their way into fish and seafood, which are then eaten by humans.

Toxic soup

In some places this toxic soup is making every link in that chain sick, including people. Exposure to these chemicals has been linked to cancers, infertility, and a whole host of disorders. Astoundingly, 92 percent of Americans have plastic and chemicals from plastic in their systems. Revealing the scale of the issue was often heart-rending for Jo's team, who saw first-hand how the ocean's animals are suffering from the rush of plastic flooding their home. Dolphins feeding on discarded shopping bags, turtles with drinking straws lodged in their nostrils, seals trapped by plastic netting. One million seabirds and 100,000 marine mammals are killed every year from plastic in the oceans.

TEN THINGS YOU NEED TO KNOW ABOUT
PLASTIC AND OUR OCEANS

* By 2050, there could be more plastic than fish in the sea.
* It takes 500 to 1,000 years for plastic to degrade.
* Each year, 11,000 pieces of microplastic are eaten by humans who consume seafood.
* Five hundred and fifty million plastic straws are used every day in the US.
* 50 percent of plastic is used once and thrown away.
* It costs just 1 cent more to make a compostable plastic bag than a traditional one.
* Only 14 percent of the world's plastic packaging is collected for recycling.
* Up to one trillion plastic bags are thrown away every year.
* The ten most common items of marine litter found on beaches worldwide are made of plastic.
* Five hundred billion plastic bottles are used worldwide every year.

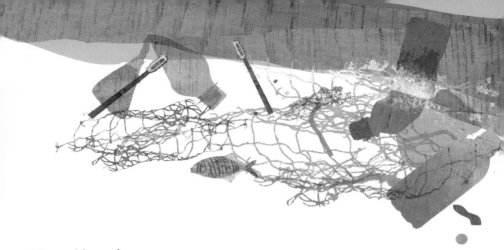

Disposable myth

Jo's determination and passion is matched by Plastic Oceans Ambassador and champion freediver Tanya Streeter, who copresented the film. Born and raised in the Cayman Islands, Tanya forged a career in the ocean, holding ten world records in freediving. In 2002, she held the overall "No Limits" freediving record—descending deeper than any other male or female—reaching 525 feet in one breath. "When I was a child, my playground was the sea. I felt cared for by the ocean—it was my safe place," recalls Tanya, who now lives in Austin, Texas. "It's a gift to be able to give something back, having had such a successful career under the waves."

"We've been told since the 1950s that plastic is 'disposable,'" says Tanya. "But it's not disposable—it's permanent and indestructible. Pretty much every piece of plastic ever made—more than eight billion tons of it—is still on the planet, leaching toxins." Things may finally be moving in the right direction, with countries such as China, France, and Morocco banning lightweight plastic bags; Costa Rica pledging to ban all single-use plastics by 2021; Kenya banning bags and introducing fines and jail sentences for plastic polluters; and nations like Germany bringing in strict new packaging and recycling laws.

But there's already enough plastic waste out there to cover a country the size of Argentina, and if plastic production increases at recent levels, these efforts could be just a drop in the ocean—more plastic has been produced in the past ten years than in the entire previous century. "It's frustrating to watch politicians obstructing laws on plastics that would make a huge difference. Given that our planet sustains our life, we should be putting her first at all costs," says Tanya. "We're at the tipping point and the madness has to stop, but at least everyone is finally acknowledging the issues."

Time to do your bit

Plastic is an environmental issue where we can all make a difference every day. "We all have power as individuals and consumers to kick this addiction," affirms Tanya. "It's amazing how much plastic each of us can prevent from entering the natural environment by making a few lifestyle changes." Jo and Tanya are out to create a wave of change, and it's one that we can ride. "We all need to rethink our plastic behavior," says Jo. "Our demand for convenience has inadvertently created a 'throwaway' lifestyle with an indestructible product. With plastic, 'away' simply doesn't exist."

Find out more about the Plastic Oceans Foundation at plasticoceans.org.

EIGHT WAYS YOU CAN HELP SAVE OUR SEAS

* If you have to use something plastic, then make sure you recycle it.
* Create a plastic-free goody bag and try to keep it with you whenever you go out. It could include items like nontoxic straws, foldable bags, utensils, and a reusable cup.
* Take your shopping home in reusable bags, avoiding plastic ones.
* Give up chewing gum—it's made of synthetic rubber, which is plastic.
* Skip the plastic bottle, drinking from reusable bottles or mugs instead.
* Don't celebrate by releasing balloons, which could end up in the ocean.
* Choose foods at the grocery store that are not wrapped in plastic.
* Eat fewer takeout and fast foods (this will be good for your health, too).

"Plastic is not disposable—it's permanent and indestructible. Pretty much every piece of plastic ever made—more than eight billion tons of it—is still on the planet, leaching toxins"

EAT, DRINK, AND REPAIR

Is your toaster broken? Camera stopped working? Zipper stuck on your favorite jeans? Visit your local, friendly Repair Café and learn how to fix what's broken.

Over the past ten years, Repair Cafés have been springing up around the world where people of all ages and backgrounds meet with the common aim of repairing things. Whether it's a malfunctioning blender or a wobbly chair, a faulty computer or a burned-out hair-dryer, a rusty bicycle or a damaged ornament, you can take a broken item to your local Repair Café and, with the help of volunteer specialists, learn how to mend it.

Tools, materials, and expertise are available at each location to enable you to make the required repairs and there is a strong community "fix-not-throw" culture as people come together to share their skills and learn new ones—all while enjoying a friendly chat over a tea or hot chocolate.

MAKE
DO AND
MEND

Repairs

- Hole in jeans
- Bicycle puncture
- Picture frame
- Watch
- Button on shirt
- School bag
- Skateboard

Why repair?

So much stuff is thrown out just because it's damaged or not working. Most of it is incinerated, increasing dangerous emissions, or ends up in toxic landfill sites. The average American produces around 4 pounds (1.8 kg) of trash every day. But a good portion of things that go in the trash could often be repaired and reused.

Going back fifty years or more, people tended to be more resourceful, learning skills to make and repair their possessions so that they lasted a lifetime. Nowadays, people are encouraged to throw out what's broken or out-of-date in favor of something shiny and new. It costs more to repair a TV or computer than it does to buy a new one, and this replace-rather-than-repair-based economy has discouraged people from trying to mend items. Thankfully, though, change is afoot. Fueled by environmental and economic concerns, more people are taking an interest in caring for and repairing their possessions.

It's surprising what can be fixed with just a little effort, knowledge, and assistance. Repairing something reduces waste, is more energy efficient than sending materials for recycling, and lowers the volume of raw materials and energy required to make new products. Do the mending yourself, and you'll save money, too. A fixing-and-making model also sends out a clear political message— that society can no longer afford throwaway habits.

Learning how to make something almost as good as new (and sometimes even better) brings a great deal of personal satisfaction and a sense of achievement. And practical skills can make a real difference to your community.

Café culture

The Repair Café movement began with sustainability journalist Martine Postma. Shocked by the amount of waste in society, she organized the first Repair Café in Amsterdam in October 2009. It was a success and, two years later, Martine went on to form the Repair Café Foundation. This nonprofit organization, based in the Netherlands, provides professional support to communities across the globe that wish to start and run their own Repair Café. Now there are more than 1,500 Repair Cafés worldwide, each striving to make the art of repairing things a part of their community. With the help of experts, they aim to bring together people from all walks of life to encourage a greater push for environmental sustainability.

The Repair Café celebrates people's practical and specialist skills and encourages them to share their expertise. It isn't somewhere you go to have an expert do an inexpensive repair on your toaster—it's a place where you're taught the skills to make your own repairs. Many who visit often stay to learn more skills and go on to help others to repair whatever's brought through the door—kettles, vacuum cleaners, and record players, to name a few.

At the same time, Repair Café encourages people to appreciate the value of what they have, and to adopt a mindset for a sustainable society. They show people that learning to fix items can be interesting, creative, and fun.

Reasons to visit a Repair Café:

* You have something that's broken that you'd like to fix rather than throw away.
* You have repair expertise—such as sewing, crafting, mechanics, electrical, or computer skills—that you could share with others.
* You're passionate about living more sustainably or are just interested in learning to fix everyday items.

Find your nearest branch at
repaircafe.org/en/

- Picture frame
- Watch
- Button on shirt
- School bag
- Skateboard

GO RETRO

Buying vintage clothing is a simple way to save clothes from landfill. Here's how to use it to create a truly individual look.

The clothes you wear can have a big influence on how you feel about yourself and how you want others to see you. By choosing a certain outfit, you can either stand out from the crowd or melt into the background. So if you're feeling loud and proud, you might want to wear something bright to get yourself noticed or something ultra stylish to create your own individual look. But it's easy to fall into the trap of buying cheap, mass-produced clothes that have a negative impact on the environment and often mean you end up looking like everyone else.

If you want to create your own personal style and are on a tight budget, then consider buying vintage. Also known as secondhand clothing, retro garments are "one-offs," so you won't turn up at a party and see someone else wearing the same thing. Fashion models Cara Delevingne and Kate Moss are known for their love of retro clothes for exactly this reason.

If you love fashion, you may have noticed that fashion goes in cycles, so what was cool thirty or forty years ago has an influence on what the fashion designers are producing now. At the moment, fashion from the 1970s and '80s is making a comeback, so all the clothes your mom or aunties wore as teenagers are suddenly on trend. If you're lucky, they may still have a few old favorites stashed away at the back of their wardrobes—ask them nicely, and they may let you borrow them.

"In order to be irreplaceable, once must always be different."

Coco Chanel

Recycling

Wearing vintage clothing is not just about standing out from the crowd, it's also about recycling clothes and avoiding piles of unwanted textiles from mounting up in landfill sites. If you've ever heard the expression "one person's trash is another person's treasure," it certainly applies to fashion. A retro jacket from the 1960s might not look good on your grandma any more so she'll be keen to throw it out, but it could look great on you. So why not invest in a few one-off items that can transform your look and be good for the environment?

Mix and match

You don't have to go full-on vintage and dress yourself from head to toe in clothes from one era (or you might end up looking like you've just stepped off the set of a PBS period drama). Instead, try adding a few vintage accessories like scarves, hats, sunglasses, or cardigans to outfits you already have in your wardrobe. A vintage jacket can suddenly transform your tired old jeans, and a simple 1960s dress can look great under a modern jacket worn with tights and chic boots.

Telling a story

Every item of vintage clothing that's been passed on over the years has its own history. Unlike a new dress that's just come off the rails, its secondhand cousin has been worn and loved by another person from another time and place, which makes it all the more interesting. Look inside the labels to find clues to the history of the items and research the label online. If you're lucky enough to stumble across a collectible designer label, you could end up with a valuable piece of clothing that you could sell on later.

Make it your own

If you're any good at sewing, it's amazing how easy it is to transform an old dress or shirt by adding some stylish buttons and, if you're really adventurous, you can buy a dress that has an interesting fabric and turn it into a new top or skirt. Also, if you see something you love but it's too big, you can easily have it altered to fit you.

WHERE TO BUY VINTAGE

❯ **Thrift stores:** Most towns have at least one thrift store selling clothes that local people have donated. Comb through the racks and you might find something special for just a few bucks.

❯ **Markets:** Many towns and cities have weekend markets where traders sell vintage clothes alongside stalls selling food and household items. Don't be put off if something looks dirty because you can wash it at home or have it dry cleaned.

❯ **Fashion fairs:** Look on your local town hall website to find out about any forthcoming fashion fairs where traders sell vintage clothing alongside new items by lesser-known designers. There may be a small entrance fee.

❯ **Online:** If you're looking for something specific, maybe a vintage piece from a certain period or something by a particular designer, search online at eBay or Etsy.

SWEET AS CAN BEE

You've probably heard that bees' numbers are declining, but how much do you really know about these important pollinators? And is there anything you can do to aid their survival?

"If the bee disappeared off the surface of the globe, then man would have only four years of life left. No more bees, no more pollination, no more plants, no more animals, no more man."

Have you heard this quote before? There is debate as to who actually predicted this gloomy scenario for the human race, and also if it is true. Although a good many crops depend on bees for pollination there are still plenty of staples—wheat, rice, and corn among them—that don't. As quotes go, though, it packs a punch, particularly as bee colonies have been declining at an alarming rate over the past decade, thanks to a toxic mix of pesticides, parasites, disease, and habitat loss. We may well keep wheat, rice, and corn, but without bees we would say goodbye to apples, avocados, onions, almonds, and broccoli. And coffee would be hugely expensive and rare.

There are up to 20,000 species of bees and 9,000 of them are to be found in North America. They have but a single source of food: nectar and pollen from plants. From this, they have developed a production line of manufactured goods: depending on type, bees produce honey, beeswax, propolis, and royal jelly as well as fitting in their day job of pollinating much of the world's flora.

Inside a colony

Bees really are terrifically hardworking. If you watch them out and about, they are always on the go. You never see a bee putting its feet up or, rather, you never see a worker bee putting its feet up.

Bees range in size from the minute stingless bee, which is less than 0.07 inch (2 mm) in length to the Megachile pluto or leafcutter bee, whose females can grow to a massive 1.5 inches (39 mm).

Some bees are classified as solitary. These are usually furrier than their honey or bumble cousins and unlike the others do not live in hives but more commonly in nests in the ground or possibly in tunnels they drill into wood, like the carpenter bee. Most people are more familiar with the honeybee, which is the guardian who pollinates more than 100 crops in the US alone.

A bee colony has three distinct castes—the queen (usually just the one), the males or drones, and sterile female workers. The drones have no other function than to mate with the queen, after which they die or are banished from the hive. The queen then engages in a frenzy of egg-laying—often up to 2,000 in a day. She can live for up to four years while her subjects ebb and flow and if a new queen appears in the hive, she will withdraw with a few of her faithful retinue to set up a new colony elsewhere.

But it is the worker bees who keep the whole enterprise on the rails. They don't have long to live—between fifteen and thirty-eight days—but they cram a great deal into this time. Worker bees clean the cells, tend to young bees, pander to the queen, build the honeycomb, forage for food, pack it away in the hive, and feed everyone. And when they are not doing all that, they dance—a bee will return to the hive and let its colleagues know where the best pollen can be found by doing a little jig, called the waggle dance. This involves shaking its middle and aligning its body in the direction of the nectar!

How to help

There has been a growing awareness recently that bee numbers are declining. This is bad news for both gardeners and agriculture. A particularly dangerous foe is the varroa mite, a parasite that is laying waste to bee colonies worldwide, but bees are also experiencing a loss of natural habitat. Under these circumstances, it would be easy to point the finger of blame at urban sprawl, but city bees are doing rather well. Instead, everyone can do their bit to aid bees by providing them with a fertile source of food. A garden helps, but pack a window box with herbs like lavender, borage, rosemary, and thyme and bees will love you for it.

Did you know?

* A honeybee visits between 50 and 100 flowers every collection trip.
* A colony consists of 20,000 to 60,000 honeybees and one queen.
* Honey has antiseptic properties and is thought to be helpful for many things, including sore throats and allergies.
* The honeybee is the only insect that produces a food eaten by humans.
* Honey lasts for ages. An explorer found a jar in an Egyptian tomb and said it still tasted great!
* A bee's wings beat 11,400 times per minute, and it's this sound that makes their famous buzz.
* A third of all the plants we eat have been pollinated by bees.

WAYS TO MAKE SURE BEES ARE HERE TO STAY

» Put up nesting boxes
Honeybee nesting boxes (not hives for honey collection) designed to help conserve wild bee populations are available from online stores and conservation charities.

» Herbs for all
The beauty about growing culinary herbs is organic plants are easy to source, even from mainstream garden centers. You can first pinch leaves for cooking and then let the plants flower for the bees. Thyme and origanum are perfect for window boxes or hanging baskets as are compact varieties of lavender and creeping varieties of rosemary.

» Garden naturally
Simply doing less garden tidying will help bees. Leave some areas of grass to grow a little longer, for example, on a sunny bank or alongside a hedge, and you will see wildflowers grow. Most of the pollen collected by honeybees in the fall comes from ivy flowers, so if you can postpone cutting back mature ivy until winter this will also help the bees.

⊗ Plant choice

As there are now fewer flowers in the countryside, gardens are an increasingly important source of nectar and pollen. There are plenty of bee-friendly plant lists available so you can improve the menu on offer and simply observing neighboring gardens to see which flowers are visited by bees in your locality will get you started. Aim to have a mix of different flowers. Raise your own plants from seed or cuttings if possible so you can grow them organically.

⊗ Go local

Could you buy honey from a local beekeeper? Is there a nature reserve in your area? Even if the latter was initially set up to conserve wildflowers and butterflies, they are also useful foraging places for bees so visit or support them if you can. Could you campaign for your local park or cemetery to be more environmentally friendly by cutting the grass less often?

⊗ The bigger picture

There is now evidence that neonicotinoid-containing pesticides, widely used in agriculture on rapeseed, in commercial horticulture for raising plants, and in some garden bug guns, can harm bees. While there are other studies that have found no correlation, you might want to support businesses that are trying not to use them.

⊗ Do your research

To find out how bees live in the wild, read Thomas D. Seeley's *Following The Wild Bees*. This will help you understand these wonderful creatures. A world authority on wild bees, Professor Seeley has been studying wild bees in Arnot Forest in New York for forty years.

IN THE PALM OF YOUR HANDS

The production of palm oil is having a devastating impact on rain forests and the animals that live there. Find out what you can do to help.

What is it?

Palm oil grows in tropical rain forests, mostly in Malaysia and Indonesia. It comes from the ripened fruit of the African oil palm, which is pulped and transformed into an edible vegetable oil. A bit like a plum, its fruit has two parts—a fleshy outer part and an inner pit. Because both of these make oil, the palm plant is far more productive than alternatives such as sunflower, coconut, canola, or soybean.

Why use it?

It has a high melting point. This means it can be added to food and products to give them a thicker, more buttery consistency, or to keep them from drying out. Animal fats can do the same thing, but they're a lot more expensive to produce.

What's the problem?

To create a palm oil plantation, the local rain forest is burned or chopped down and the native wildlife, including orangutans, tigers, and elephants, are killed, captured, or driven away. These fires release high levels of CO_2 and have been blamed for extreme air pollution in neighboring countries.

What's it used in?

Palm oil is said to be found in half of all the products you see at the grocery store, from food to cleaners to cosmetics, including margarine, cookies, pizza dough, donuts, chocolate, ice cream, popcorn, soap, toothpaste, shampoo, lipstick, chips and packaged bread. It's even used in baby formula and dog food. You might not always see it on the food labels, however, as it's often described in generic terms, such as vegetable oil or vegetable fat.

Lost habitat

The demand for palm oil is driving uncontrolled deforestation in the countries that produce it. The World Wildlife Fund (WWF) estimates that every hour about 300 football fields of forest are cleared to make way for palm oil plantations. Its 2018 *The Living Planet Report* highlights that wildlife has decreased by 60 percent worldwide since 1970 and that the habitats of the critically endangered pygmy elephants, Sumatran rhinos, sun bears, orangutans, tigers, and proboscis monkeys—all native to Malaysia and Indonesia—have been destroyed. It added that Sumatra had already lost 12.5 million hectares of natural forest between 1985 and 2010. What's more, demand for palm oil is expected to double by 2030 and triple by 2050. Without drastic change, the Sumatran rain forest could disappear within just twenty years.

Should you stop buying products that contain it?

You might think we can all make a difference by giving up products that contain palm oil, but this isn't necessarily the case. Some people say that avoiding it could make the situation more difficult by forcing producers to grow other crops that are even worse for the environment. Also, the palm oil industry currently offers jobs to people living in poverty in developing countries. Instead of cutting it out altogether, they argue that palm oil, which needs less land to grow than other vegetable oils, should be grown more responsibly and in a sustainable manner. There are two ways you can help to make sure this happens:

* **Look for the label:** A group called the Roundtable on Sustainable Palm Oil (RSPO) formed in 2003 to try and get those in the industry to work together to stop palm oil production from damaging the planet. There's now an RSPO label that certifies products that have been made with sustainable palm oil. Look for this when you're shopping at the grocery store.
* **Call the company:** If there's a product you want to buy that contains palm oil but doesn't currently have the RSPO label, call or write to the company and urge them to use certified sustainable palm oil instead. Most packaged products are labeled with their companies' contact information. Ask them to take action to support a more responsible palm oil industry.

Which brands are using unsustainable palm oil?

Some product manufacturers have said they'll only buy palm oil that they know has been produced in a sustainable way. But many brands say it's difficult to know for sure whether or not farmers are actually following the rules. A recent report released by Greenpeace named twelve companies in particular that continue to use palm oil from the twenty suppliers most responsible for destroying the habitat of orangutans. You can find out more about these companies and the products they produce at greenpeace.org/usa/.

Together, we can insist that those who grow and make palm oil use processes that protect natural animal habitats, foster healthy ecosystems, and create positive opportunities for local communities. Maybe then the planet's long-term future will stand a chance.

To learn more about the impact of the palm oil industry, go to worldwildlife.org.

HIDDEN INGREDIENT

Many products that use palm oil aren't clearly labeled. There are in fact more than 200 terms for palm oil, including vegetable oil, vegetable fat, palm kernel, palm kernel oil, palm fruit oil, palmate, palmitate, palmolein, glyceryl, stearate, stearic acid, elaeis guineensis, palmitic acid, palm stearine, palmitoyl oxostearamide, palmitoyl tetrapeptide-3, sodium laureth sulfate, sodium lauryl sulfate, sodium kernelate, sodium palm kernelate, sodium lauryl lactylate/sulfate, hydrated palm glycerides, etyl palmitate, octyl palmitate and palmityl alcohol.

GOING GREEN

Vegetarian, vegan, flexitarian—there's a diet for everyone. And one that's been talked about a lot recently is the "plant-based" option. If you've heard of it, you may have thought this was the same as being a vegetarian or even a vegan. But there are differences.

Plant-based diets are growing in popularity and depending on the type you choose, it may or may not be meat-free.

* **Vegan:** No meat, egg, dairy, or any animal product. But this doesn't necessarily mean you only eat veggies. You could choose to eat nothing but chips and dark chocolate.
* **Vegetarian:** Meat- and fish-free, but it doesn't have to be based on vegetables and fruit. Someone who eats fish and vegetables but refrains from eating meat is called a pescetarian.
* **Plant-based:** Focuses on eating whole fruits and vegetables, consuming lots of whole grains, and minimizing (or cutting out, depending on which plant-based diet you choose) animal products and processed foods like cereals, white bread, canned vegetables, cookies, cakes, and soft drinks.

There are also crossovers with another new style of diet that has recently emerged—the "flexitarian." Instead of following a strict vegetarian or vegan diet, many people are now choosing to eat mainly plant-based foods but still including meat and/or fish occasionally.

    A well-planned plant-based diet can be both nutritious and healthy, but whatever you choose to eat, don't feel pressured into labeling yourself, or to follow a specific plan just because it's fashionable. Eat in accordance with your beliefs and preferences, irrespective of other peoples' opinions.

THE NUTRIENTS YOU NEED

If you like the sound of a plant-based diet, there are six main nutrients to be aware of:

◎ Protein
Plant-based sources of protein include beans, lentils, chickpeas, soy, nuts, and seeds, as well as wheat, rice, and corn. Milk, yogurt, and cheese also provide protein, in addition to meat-alternatives such as mycoprotein. Plant-based sources of protein are generally incomplete (they don't contain all of the essential amino acids, which are the building blocks of protein), meaning it's essential to eat a variety of them every day. Soy, quinoa, and hemp are the only plant-based complete sources of protein.

◎ Iron
Iron is responsible for making red blood cells, and red meat is the most easily absorbed source. Good plant-based sources of iron include legumes; dried fruit; dark green vegetables such as watercress, broccoli, and spring greens; whole-grain bread and fortified breakfast cereals. Try to include a source of vitamin C with your iron source to help you absorb it better.

Calcium

Dairy foods are rich in calcium, which is essential for healthy bones and teeth. Plant-based sources of calcium include calcium-fortified dairy-free milks, figs, almonds, green leafy vegetables including kale and bok choi, kidney beans, sesame seeds, and tofu. Although spinach does contain calcium, it isn't easily absorbed.

Vitamin B$_{12}$

Other than meat, milk, eggs, and cheese, sources of vitamin B$_{12}$ include fortified yeast extracts and fortified breakfast cereals, as well as soy products. A vitamin B$_{12}$ deficiency can lead to low energy levels and nerve damage, which is why you may want to consider taking a supplement containing at least 10 micrograms of vitamin B$_{12}$ a day, or 2,000 micrograms weekly (if you're not eating at least 3 micrograms of vitamin B$_{12}$ from fortified foods).

Selenium

In addition to meat and fish, nuts—in particular Brazil nuts—are a good source of selenium. Just two to three Brazil nuts a day provide your body with the daily amount of selenium that it needs.

Omega-3

Omega-3 fatty acids are essential for your heart health, and oily fish is one of the richest sources. Ideally, you should eat two portions of fish a week, one of which is oily. If you don't eat fish, it's advised to include flaxseed (linseed) oil, canola oil, and soy-based foods such as tofu and walnuts, as well as eggs enriched with omega-3, on a regular basis.

And remember:

- * Base your diet on a wide variety of plant-based foods including fruits, vegetables, lentils, legumes, nuts, and seeds—eat a rainbow.
- * Choose higher fiber or whole-grain starchy foods such as oats, whole-grain bread, whole-grain pasta, and brown rice rather than refined varieties.
- * Include heart-healthy fats such as olive oil and canola oil for cooking and for salad dressings.
- * Reduce your intake of sugary drinks as well as foods that are high in calories and low in nutrients such as cakes, cookies, and candy (have these only occasionally).
- * Make sure your multivitamin includes (in addition to vitamin B_{12} and calcium) iodine and vitamin D. You might also consider taking a supplement containing at least 10 micrograms of vitamin D a day, especially in fall and winter.
- * Drink at least eight glasses of fluid a day—water is best.
- * Eating a more plant-based diet is good for the environment and can have a positive effect on your health, too. There is, however, no nutritional need to cut out foods such as meat, poultry, and dairy, especially if you enjoy them!

BECOME A NATURE GUARDIAN

The natural world needs your help. There are many ways you can show you care about your local environment, and they may be more enjoyable and rewarding than you think.

The destruction of natural habitats, animal neglect, and the rise in pollution are just some of the important issues facing the planet today. If you're someone who cares for nature and the environment, you might be wondering what you can do to help. One answer is to give a little time and energy to support animal welfare or environmental causes in your community. You'll help to make a difference and gain experiences to remember at the same time.

Here are some suggestions on how to get involved:

🡢 Plant trees

They're important habitats for birds and wildlife and vital for the health of the land—they help to prevent flooding and maintain the purity of the air we breathe. With the increase in deforestation in many parts of the world and with trees being lost to neglect, disease, age, and cutting, it's important to plant more trees. There is a saying: for every tree lost, plant three more. Take a look at One Tree Planted (onetreeplanted.org) and the National Forest Foundation (nationalforests.org) on how to support tree planting in your area.

Care for the birds

Many wonderful species, including woodpeckers, condors, cranes, warblers, and jays are, according to the National Audubon Society, in big trouble in the US. To prevent a further decline in numbers, action is needed to look after those that remain. To find out how you can get involved, check out the Audubon Society (audubon.org), the American Bird Conservancy (abcbirds.org), or look for local charities in your area.

Tidy-up natural habitats

Littering and the illegal dumping of waste can affect the coast, rivers, local parks, wildlife areas, and even the streets around your home. Some items that do not biodegrade, such as certain types of plastic, pollute land and water and cause problems for animals and wildlife. Consider volunteering with a community group where you live to help tidy up your area and create a better environment on your doorstep. It's amazing how cleaning up an area can transform it, and it can help foster a healthy respect for these vital green spaces in your community, too.

Become a conservation volunteer

People love to explore the countryside, but with so many visitors as well as the problems of natural erosion, natural attractions require constant care and repair. For example, pathways, fences, stone walls, and hedges in popular parks and conservation areas regularly need attention. You can play a vital role in looking after the natural environment by becoming a conservation volunteer. You will learn how to undertake a range of activities in both urban and rural areas, from planting trees and wildflower meadows, to building stiles, clearing footpaths, or creating food growing projects. Look for opportunities in your local area or contact the National Park Service: nps.gov/getinvolved/volunteer.htm.

⊙ Volunteer at a horse sanctuary

Love horses? There are many equine rescue centers caring for neglected, ill, or elderly horses, ponies, and donkeys that need hands-on volunteers to help with horse care. Take a look at Habitat For Horses (habitatforhorses.org/volunteer/) and U.S. Horse Welfare & Rescue (habitatforhorses.org/volunteer/).

⊙ Support your local wildlife organizations

Most small local organizations rely on a large number of regular and occasional volunteers. These grassroots movements work to make local areas wilder and ensure that nature is part of life for everyone. Consider how you could get involved with community wilding projects through local fundraising, using your free time to give practical support.

WONDER WALK

Next time you head outdoors, why not try this simple exercise?

Walk at a natural pace, breathing fully and deeply, and pay attention to every step you take. Place your hands wherever it's most comfortable—by your sides, on your belly, or behind your back.

Notice the subtle movement in your legs and be aware of how your feet touch the ground and keep you upright. Feel how each foot swings ahead in turn and how the heel hits the ground—then the ball, then the toes.

Slowly shift your awareness to the sights and sounds around you. Engage your senses and take a deep breath in through your nose. What can you smell? Notice your surroundings—the light of the sky, the shapes of drifting clouds, the colors of the leaves.

With each thought, simply notice it, and then let it go. Note how you're feeling in yourself at this moment. Has your breathing changed while you've been walking?

JUST ADD SUNSHINE

Anyone can grow plants from seed. The key thing, as with any nurturing activity, is patience and a little tender loving care.

Growing plants from seed rather than just buying them in pots can be amazingly fulfilling, especially if you're using clean recycled items and organic methods rather than commercially raised plants with their associated fossil fuels, plastics, and pesticides. What's more, there's a wider range of plants available to grow from seed compared to buying potted plants—and the more varied the range of plants grown, the better for biodiversity (and the planet) as a whole.

Source your own

There's no shortage of places to buy seeds, but if you're a beginner then scouring a garden center shelf can be bewildering. Not only is there so much choice, there's a fair bit of jargon, too. Instead, ask gardening friends or relatives if you can rummage through their seed cans. Most keen gardeners will have more seed than they'll ever get around to sowing.

Another approach is to look out for seed swaps in your area, usually organized by community groups or gardening societies. You simply show up and swap or buy a few packets.

TIME TO GROW

▶ It's okay if some don't make it
Seed sowing is not without its frustrations. Tiny little seedlings are prone to drying out, drowning, or getting fried in the sun. Keep in mind the bigger picture. You're coaxing life out of the seed by providing warmth and moisture. You're triggering a process of growth and then photosynthesis that will enable the plant to grow and complete its life cycle. Mother Nature has provided far more seed than she expects to grow, so some failures are natural.

▶ Start small
Aim to nurture and grow a few plants initially rather than starting off with lots of different seeds and risking becoming overwhelmed with caring for them all at once. After all, particularly on an indoor windowsill, each pot of seeds sown will need to be transplanted into individual pots so space can soon run out.

▶ Light them up
While warmth and moisture are the triggers for germination, sometimes light is required too (check the seed packet). Once seedlings form, they'll need light for sturdy growth, so if you're keeping pots on a windowsill, you'll need to rotate them daily for even growth.

▶ Watch out for fungi
Make sure pots are clean and you have some fresh seed compost. Warmth and moisture will also encourage bacteria and fungi to grow, which can infect the seedlings, so keep everything as clean as possible right from the start.

▶ Be the seed
You need to anticipate your seeds' needs. The aim is to keep the seedlings growing steadily into young plants. If they dry out or get crowded, they may not recover. Check to see if there's a wilting leaf or a chill behind the blinds in the evening. Place a finger on the compost to see if it's dry. If so, they need a drink.

A STEP-BY-STEP GUIDE

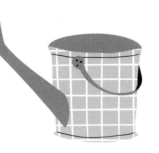

*You don't need a greenhouse to grow plants from seed.
Those that need a bit of extra warmth can start off indoors
on a windowsill; others can be sown straight into the ground
from mid-spring onwards.*

What to grow

Pretty annual flowers, which last only one season, include orange calendula,
red-flowered nasturtium, blue-flowered nigella with its feathery foliage, and
fragrant climbing sweet peas.

Edibles to grow from seed include salad leaves such as peppery arugula and
sweet, buttery lettuce, as well as herbs such as pungent cilantro, fast-growing dill,
and flavorful basil.

If you want to grow vegetables, consider climbing beans and quick-to-crop
radish. Bush beans and dwarf tomatoes are well suited to small spaces.

To sow seeds indoors, you'll need:

* Clean small pots (3 inches/7 centimeters in diameter) with drainage holes
* A tray or saucer to hold the pots
* Clear plastic bags and rubber bands
* Labels and a waterproof pen
* Fresh seed-starting (or multipurpose) compost

1. Fill the pot with compost. Press the surface firmly so it's 1 inch
   (2.5 cm) below the rim.
2. Water gently from below—place the pot in a tray of water to
   moisten the compost. Drain any excess.
3. Sow a pinch of small seeds over the compost surface—large seeds
   can be pressed in. Scatter loose compost over the seeds.
4. Label the pot with the type of seed and the date sown. Place in a
   plastic bag and put a rubber band around the top.
5. Wait.
6. When a seed germinates, the first pair of leaves are part of the seed.
   The next pair are the true leaves. When at least one pair of true leaves
   has opened, move the seedlings into new pots so each plant will have
   its own rooting space.

FORGOTTEN WAYS

Old-fashioned traditions and practices have a lot to teach you about taking better care of yourself and the planet in the here and now.

Ever wondered what people did with their time before the invention of computers, cell phones, the internet, and even television? You might think life must have been boring, but most people kept busy practicing simple arts and crafts, which were part of everyday life. In the spirit of revival, holidays are a great opportunity to discover these traditional and almost forgotten crafts, and enjoy the experience of acquiring some practical skills.

At first glance, learning something "old school" doesn't seem that enticing. What's so wonderful about baking bread, making soap, or constructing a birdhouse? With the growing interest in self-sufficiency, upcycling, making, and repairing, along with the joy of being completely immersed in a creative project, more people are recognizing the value of gaining natural, useful skills that create less waste, make a difference, and save you some money. Plus, it feels rewarding to be able to make something and it's a lovely way to fill your free time.

PROJECTS TO TRY

⟩ Press flowers

The art of pressing flowers has long been practiced in Asia and became popular in the West during the Victorian period. It involves drying and flattening flowers (and leaves) in a flower press to preserve them so that they can be used for art projects or as a souvenir. Choose fresh blooms from a florist or, if you have a garden, ask your parents for permission to use a couple of flowers for your project. To prevent browning, the flowers need to be dried quickly. If you don't have a flower press, you can arrange your blooms between paper or cardboard and then place a heavy book on top of them. Remember to change the sheets of paper after a few days to remove moisture, but be careful not to spoil the arrangement. After a few weeks, the delicate flowers will be completely dry and, with care, can last a lifetime.

⟩ Make lemon ice pops

Stay cool this summer by quenching your thirst on a refreshing homemade lemon ice pop. These are super-easy to make. All you need is one cup of freshly squeezed lemon juice, two cups of water, and a spoonful of honey for sweetness. Combine the mixture in a pitcher and pour into ice pop molds. Freeze until solid.

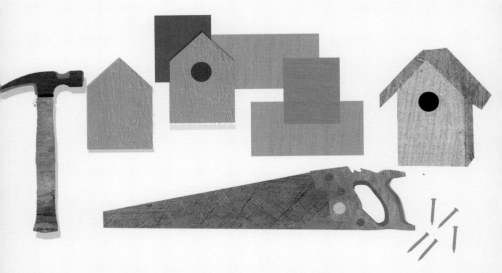

❯ Make lavender pillows

Since ancient Egypt, lavender has been used to ward off infections, perfume clothes, and repel insects, and is known for its antibacterial, antiseptic, and calming properties. This flowering herb with its dreamy aroma offers a perfect drawer freshener for your bedroom. To make your own mini lavender pillow, you'll need a pack of dried lavender, two squares of fabric, scissors, needle, and thread. Cut the fabric into 4-inch (10.2 cm) squares. Align the two pieces of fabric and sew around the edges on three and a half sides. Turn the fabric sachet inside out and fill it with lavender. Carefully sew up the gap to secure the contents. You can add a frill around the edges or decorate the pillow with buttons or sequins.

❯ Make a quill pen

Made from the feathers of large birds, quill pens were popular around the fifteenth century and were the writing tools of the day. To make your own, find a suitable flight feather that's been naturally shed by a crow, goose, or turkey. The feather should be just the right length to sit comfortably in your hand with a point that's sturdy enough to hold ink. Trim away the lower feathers so that you can grip the quill pen with ease. Carefully cut the end of the feather at an angle to create a V-shaped point, like the nib of a fountain pen. Dip the quill point in a pot of ink and start writing.

◉ Homemade washing powder

Before there were hundreds of cleaning sprays, powders, and solutions in supermarkets, it was necessary for people to make their own, often from everyday products. DIY detergents have the advantage of producing less chemical-laden waste water, and they don't expel additives. The core ingredients of any DIY detergent are much the same: a bar of natural soap (grated), 8.5 fl oz borax, and/ or the same amount of washing soda. All of these should be available from your local store, and certainly online. For those a little nervous about using borax (a natural mineral with a sinister-sounding name), it is quite possible to double up on washing soda instead. If you are in favor of sweeter-smelling socks, add a few drops of essential oil or pick a fragrant soap to grate in. And if you're looking to boost your stain-busting power, there is the option of adding white vinegar and lemon juice, both of which help to cut through grease.

MAKE IT NEW, MAKE IT YOU

Upcycling old clothes, magazines, and more is a great way to create something new, personal, and unique out of stuff you might otherwise throw away.

The world produces over 2 billion tons of municipal solid waste every year—that's enough to fill more than 800,000 Olympic-sized swimming pools. Municipal solid waste means garbage that's collected by local authorities from homes and businesses. While 2 billion tons of this is produced each year, research firms estimate only 16 percent of it is recycled and almost half is disposed of unsustainably.

Every person on this planet needs to think hard about what they consume and what they throw away—but it can be hard to start when it's so easy and enjoyable to buy new things. The good news is that your once-loved-but-worn-out stuff could have a new lease of life if you upcycle it into something that is new, different, unique, and completely personal to you.

Everything that's yours has a story: the favorite jeans you wore to that awesome party and are now too ripped to wear anymore, the comics you still kind of love but feel too old for now, the T-shirt that's gone a bit too saggy. They could just go in the trash—or you could use them as raw materials to create something new and brighten up your room or wardrobe. None of these ideas require particularly specialist skills—if there's something you haven't tried before you can easily learn how online. It's time to fight waste and start crafting.

⬢ Worn-out T-shirts

Got an old T-shirt you love but that's so ragged it can't be worn anymore? Why not turn it into yarn and crochet or knit it into something new? Starting at the bottom, cut the T-shirt into a narrow spiral ribbon no more than ½ inch (1.3 cm) wide. Once you've got to the neck, do the same for the sleeves. Wind this ribbon into a ball and you're ready to start stitching—you could make a bowl, a hat, a wall hanging, or even a bag. If you don't know how to crochet or knit yet, you can find plenty of tips and tutorials online.

⬢ Fabric scraps

If you've got old clothes you love but don't want to wear anymore, or if you've collected scraps of beautiful fabric but don't know what to do with them, why not sew them together in a patchwork and use them to cover a bulletin board? Once you've got a piece of fabric that's big enough to cover your bulletin board and stretch a little way over the back, simply use a staple gun to fix it in place. Add criss-crossed elastic ribbons, also staple-gunned to the back, to hold photos and other mementoes without getting pinholes in them.

⊗ Old paper, pictures, and other bits and pieces

It can be hard to let go of old comics, magazines, and bits of fabric—but they also take up a lot of space. Découpage is a great way to use any saved cutout bits of paper, material, and even tissue paper to create something beautiful. Simply cut out the pieces you want, arrange them over a table top, drawer fronts, an old trunk, a lampshade, cardboard cutout letters, or anything else you like, glue them in place, and then, once the glue is dry, varnish over them. If you use a special découpage glue available from craft stores, you can do it all in one step with no need to varnish afterward.

⊗ Ripped jeans

Denim is a fabulous fabric, but jeans can get ripped in places that mean you just don't want to wear them anymore. The great news is you can upcycle your old favorites into something new and exciting. Make a skirt by slitting the seams with sewing scissors (careful—don't cut yourself!), and try on the opened-up ex-jeans. Mark a point a few inches longer than you want the final skirt to be, then cut the legs off there. Sew the offcuts into the gaps between the opened legs and you will have a skirt—then simply sew up a hem. Looking for a maxi skirt? Use a different pair of jeans or even a contrasting fabric to fill in the gaps between the legs and you'll have a completely new garment. Alternatively, why not make them into a bag? If you can use the top back section, you'll have a couple of useful pockets already in place. Use a ribbon or a strip of scrap fabric to make a handle.

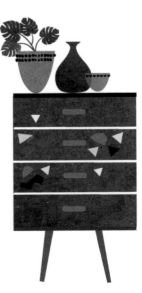

Holey sweater

Visible mending is a hot trend right now—Japanese sashiko embroidery techniques and Swiss darning are great ways to both mend damaged garments and add a personal touch. If your favorite sweater has been attacked by moths or has come by some other disaster, why not use a contrasting yarn to darn the hole, then embroider over the mending with your initials or a pretty motif?

Sometimes sweaters give way completely at the elbows, leaving the body intact. In this case you could repurpose your sweater as a knitted cushion cover. Use a sewing machine to sew together a square each from the back and front—you may wish to run a line of zigzag stitches all around to guard against fraying. Then cut outside those stitches, add in a zipper or buttons, pop in a cushion pad, and you're ready to display it.

Old boxes

Old wooden boxes can be incredibly versatile: you can paint them, upholster them, even screw them to the wall to create stylish shelves. Cardboard boxes can look great too—why not découpage an old shoebox to store your favorite mementoes? Or paint or cover a toy or storage box you've grown out of to suit your own style?

#STAYSTRONG

With all the negative news circulating about climate change and the future of our planet, it is easy to feel discouraged about the causes you believe in or are campaigning for. Being "green" can seem difficult and you may find you are harsh with yourself. Sometimes you might feel helpless, insignificant, or as if your actions don't matter. Other people may question or even mock your beliefs and the issues you support. Well, we are here to tell you that your actions do matter and the issues you support are important. Here are some things to remember when you start out on your eco-warrior journey that will help you to stay strong . . .

⊙ You don't need all the answers

What you're doing is a small part of a much larger effort. If someone challenges you about whether what you are doing will make a positive difference, it's okay to say to them: "I don't know." Remind them that you're doing what you can and encourage them to do the same. While there's a lot of negative news, there are also a lot of positive stories. Use these and your optimism to combat any negativity you face. Lead the way by setting an example and make a difference on social media, such as sharing the causes you support, the food you eat, or the clothes you wear. Showing rather than telling can have more of an impact.

⊙ You don't have to do it alone

Sometimes wanting to make a change to the world may feel overwhelming—but you're not alone. If you worry that your actions are too small to make any difference, it could be because you're too focused on trying to do things alone. If this is you, consider how you could work with others. Is there a local environmental awareness group you could join? Or why not look online and join an organization you support? That way, you won't feel so daunted. Having like-minded people to talk to will give you a boost and remind you why you care and want to make a difference. Individual effort is good—but you can do even more as part of a team.

Be happy with your effort

You're making a difference, and you should be proud of your contribution. Sometimes you may feel you don't do enough and could do more in comparison to others, but nobody is perfect. For example, you may not be vegan, be unable to cycle to college, or forget to turn your appliances off from time to time. But you don't have to be flawless—you just have to do the best you can. Maybe you've limited your use of plastics, joined a beach litter-picking group, or are paying more attention to sustainable clothing. Sometimes people just have to make the best choices they can. Celebrate every tiny change, because all those small actions add up.

Don't be scared to be heard

Some people believe that only seasoned eco-warriors should speak out about issues, but remember—everyone has a right to have their voice heard, and that includes you. You don't need to know everything, just be enthusiastic and passionate about your cause. Everyone has their own opinion and not everyone will agree, but you should be heard without fear of trying to prove something to anyone else. Having your voice heard is a form of activism and a way to campaign for an issue.

⊙ It's never too late to start
Your parents may have been activists and taken you on marches before you could walk. You may never have eaten meat or given it up years ago. Whether these are true for you or not, it's never too late to start. Don't be put off by others who think they're more environmentally active than you just because they've been doing it for longer. Think of the simple ways you could make changes, whether it's swapping car rides for public transport, having meat-free Mondays, or considering what clothes you buy. It doesn't matter when you start making a difference, as long as you start now.

⊙ Stay positive
Your influence on others is important. While it's okay to have challenging days, consider the bigger picture. Not everyone will understand or listen—but others will. Without realizing it, you've probably inspired friends or family to make changes already. Stay strong—you've got this!

PEOPLE POWER

It's easy to feel defeated in the face of scary global warming statistics, but by joining hands with fellow citizens across the world there's every chance you can have a positive impact.

When it comes to climate change, the headlines would have you believe the end is nigh. If it's not photos of bushfires, it's time-lapse videos of melting ice caps or memes of oil-drenched birds foraging through plastic. You'd be forgiven for losing faith and wondering if it's all too late. The scale of the task ahead can sometimes feel overwhelming.

But don't panic—and please come out of your doomsday bunker—for positive progress *is* being made when it comes to global warming and sustainability. Grassroots movements are bolstering international outrage and governments are listening. (Well, some of them anyway.) Awareness and acceptance is leading to action, and it's integral that news of these victories is shared as much as, if not more than, all the doom and gloom.

Climate activist Bill McKibben is the cofounder of 350.org, a planet-wide movement that wants to end the use of fossil fuels, and he believes people power can influence government policy. To date, 350.org campaigns have stopped fracking in hundreds of cities in Brazil and Argentina; compelled many universities, foundations, and churches to divest from fossil fuels; and given some of the poorest communities on Earth a voice.

"We can't just be individuals, we need to join together and be a movement," says Bill. "The best antidote to feeling powerless is activism. It doesn't make you less sad, but adds hope, solidarity, and love." He's right. Activism transforms defeatism and hopelessness into a collective force for creating a better world. When like-minded people come together to protest peacefully, they become empowered. At the very least, they raise awareness of their cause. During the weeks of Extinction Rebellion's protests, for example, there was a five-fold increase in online searches for climate change. The effects of campaigning with others who share your views go way beyond that, however, and are felt at all levels—personal, local, national, even international.

"Never doubt that a small group of thoughtful, committed citizens can change the world; indeed, it's the only thing that ever has."

MARGARET MEAD

A real difference

On a personal level, activism brings a sense of belonging, which can help to boost well-being. Acting with purpose, having a cause you believe in—picking up litter or taking all nonrecyclable packing to the grocery store to register your discontent—can also make you feel you are doing something constructive, which is one way to increase self-esteem.

Beyond that come the social benefits to the community. Every week, food redistribution charity FoodCycle serves a hot meal to more than 1,400 people who might otherwise go hungry. It achieves this by using surplus goods collected from grocery stores. It was partly pressure from consumers that led to grocery stores agreeing to donate produce they can't sell to FoodCycle and other similar organizations and also to pledge to reduce their waste. Similarly, shoppers taking back all nonrecyclable plastic to stores has helped convince retailers to sell more loose produce.

Never underestimate your ability to help shape the future. And don't think it has to be on a huge scale—it's not all about marching on Washington or joining demonstrations on the street. It can be as simple as pledging to consume less, recycle more, and make informed choices.

Activism is often the only route open to the public, to the billions of us who are not politicians or heads of global corporations. Our way of exerting influence is to stand up and be counted. As author Margaret Mead said: "Never doubt that a small group of thoughtful, committed citizens can change the world; indeed, it's the only thing that ever has."

TURNING THE TIDE

These examples show how joining forces with others can bring about positive change:

* Concerns over food waste are causing government and retailers to take action. April 2019 was named "Winning on Reducing Food Waste" month as the U.S. Department of Agriculture, Environmental Protection Agency and the Food and Drug Administration teamed up to reduce food waste. Three major grocery chains—Kroger, Walmart, and Ahold Delhaize USA—have publicly committed to achieve zero food waste by 2025 and have prioritized tracking food waste and prevention.

* Thousands of people across the US have joined the International Coastal Cleanup where local communities collect and document the trash littering their coastlines. The Cleanup began thirty years ago with just two individuals, Linda Maraniss and Kathy O'Hara, and is now an annual international event. In 2018, 35,500 people took part in the Big Spring Beach Clean in the UK, organized by Surfers Against Sewage, removing around seventy tons of plastic and litter from 571 UK beaches.

* The nonprofit organization As You Sow has engaged major brands such as McDonalds, Starbucks, PepsiCo, Walmart, Procter & Gamble, and Campbell Soup to phase out their use of polystyrene foam-based packaging and to recycle more of their packaging.

* In 2016, City to Sea, a group aiming to prevent plastic pollution at source, organized a petition with 38Degrees urging UK retailers to switch from plastic-stick cotton swabs to paper ones. More than 150,000 people signed, prompting major retailers to get on board. "We're delighted with the commitment from so many major supermarkets to 'Switch the Stick' from plastic to paper stem buds," said City to Sea's founder Natalie Fee. "While they still shouldn't be flushed, this move will stop millions of plastic stems ending up in the marine environment."

STERLING
New York

An Imprint of Sterling Publishing Co., Inc.
122 Fifth Avenue
New York, NY 10011

ISBN 978-1-4549-4299-3

Distributed in Canada by Sterling Publishing Co., Inc.
c/o Canadian Manda Group, 664 Annette Street
Toronto, Ontario M6S 2C8, Canada

For information about custom editions, special sales, and premium and corporate purchases, please contact Sterling Special Sales at 800-805-5489 or specialsales@sterlingpublishing.com.

Manufactured in Singapore

2 4 6 8 10 9 7 5 3 1

sterlingpublishing.com

Editorial: Susie Duff, Catherine Kielthy, Jane Roe
Design by Jo Chapman
Publisher: Jonathan Grogan

Words credits: Christine Boggis, Jenny Cockle, Lorna Cowan, Liz Dobbs, Donna Findlay, Anne Guillot, Lauren Jarvis, Nicola Ludlum-Raine, Lola Méndez, Victoria Pickett, Rachel Roberts, Carol Anne Strange, Ruby Etwaria Sweetman, Jo Usmar

Illustrations: Shutterstock.com, Agnese Bicocchi, Jen Boehler, Teresa Arroyo Corcobado, Trina Dalziel, Amber Davenport, Nicola Ferrarese, Beatrix Hatcher, Claire van Heukelom, Jen Leem-Bruggen, Laura Moyer, Lou Baker Smith, Sara Thielker, Ellice Weaver, Nicola Youd

Cover illustration: Laura Moyer